Elite German Divisions in World War II

Elite German Divisions in World War II

Waffen-SS • Fallschirmjäger • Mountain Troops

WERNER HAUPT

Translated from the German by David Johnston

Schiffer Military History
Atglen, PA

This book was originally published under the title,
Deutsche Spezialdivisionen 1935-1945,
by Podzun-Pallas Verlag GmbH, Wölfersheim-Berstadt, 1995.

Book design by Robert Biondi.

Library of Congress Catalog Number: 2001086977.

Printed in China.
ISBN: 0-7643-1432-7

We are always looking for people to write books on new and related subjects. If you have an idea for a book, please contact us at the address below.

Published by Schiffer Publishing Ltd.
4880 Lower Valley Road
Atglen, PA 19310
Phone: (610) 593-1777
FAX: (610) 593-2002
E-mail: Schifferbk@aol.com.
Visit our web site at: www.schifferbooks.com
Please write for a free catalog.
This book may be purchased from the publisher.
Please include $3.95 postage.
Try your bookstore first.

In Europe, Schiffer books are distributed by:
Bushwood Books
6 Marksbury Ave.
Kew Gardens
Surrey TW9 4JF
England
Phone: 44 (0)208 392-8585
FAX: 44 (0)208 392-9876
E-mail: Bushwd@aol.com.
Free postage in the UK. Europe: air mail at cost.
Try your bookstore first.

Contents

Introduction

The army, air force and *Waffen-SS* included division-strength units which filled specialized roles, at least until the end of 1941.

The units described in this book were formed and trained for specialized roles.

The mountain troops were trained and armed for combat in mountainous terrain; the parachute troops were trained and armed for parachute operations behind enemy lines; the *Waffen-SS* was trained and armed for its mission of protecting personalities in and outside of Germany.

The term "special divisions" is entirely applicable only for the prewar period and early war years. Not until later in the Second World War were these units committed as conventional units.

Werner Haupt
Waiblingen, summer 1995

1

Mountain Troops
(Gebirgstruppe)

History to the Outbreak of War

During the years of peace prior to the First World War the German Army did not have any units specially trained for mountain warfare. Not until the Imperial Austrian Army requested support in its defense of the Tyrolean Alps did the Bavarian Army create the so-called "Alpine Corps" under the command of the later *Generalleutnant* Ritter von Tutschek. In the beginning the corps consisted of the Bavarian Infantry-Bodyguard Regiment and the Bavarian 1st Light Infantry Division. The Bavarian Mountain Artillery Battalion and the Mountain Machine-Gun Battalion were created at a later date.

In autumn 1915 the Royal Württemberg Army formed the Württemberg Mountain Battalion under the command of *Major* Sproesser. The battalion consisted of six companies. Later in the war, when the front extended into the Balkans, various contingents contributed troops to mountain artillery regiments, machine-gun battalions and other formations, including motor pool battalions.

These formations were used with success, mainly in the Balkans and in the Italian Theater. Two young officers who rose to the rank of *Feldmarschall* in the Second World War earned the *Pour-le-mérite* while serving with the mountain troops: Schörner of Bavaria and Rommel of Württemberg.

The Versailles Treaty forbade the German Army of the Weimar Republic from possessing mountain troops. At Mittenwald in spring 1935, following the introduction of military sovereignty, the new German Armed Services (*Wehrmacht*) equipped and trained its first mountain regiment, the *Infanterieregiment 98*.

Other mountain units followed. The *1. Gebirgsbrigade* under the command of *Oberst* Kübler was formed at Garmisch-Partenkirchen on 1 October 1936. Kübler, formerly of the Bavarian *19. Infanterieregiment*, had previously worked on the formation of a mountain corps while serving in the *Reichswehrministerium* (Ministry of Defense). When the *1. Gebirgsdivision* was formed on 1 March 1938 he assumed command with the rank of *Generalmajor*. (*General* Kübler and his brother were both murdered in Yugoslavia in 1947.)

The new *Gebirgsdivision* (mountain division) possessed the same constituent units as an infantry division. It differed only in the organization and equipment of its divisional units plus some minor differences in uniform. For example, instead of light and heavy field howitzers, the mountain artillery regiment was equipped with the 75-mm *Gebirgsgeschütz 15*, which in 1938 was replaced by the 75-mm *Gebirgsgeschütz 36* (built by Rheinmetall-Borsig). Introduced later was the 105-mm *Gebirgshaubitze 40* which was developed and built by the Austrian firm of Böhler. The guns and other heavy weapons could be broken down and transported on pack animals.

The new division soon saw its first action during the occupation of Austria in March 1938. Two motorized detachments from the division were the first units of the German Army to reach the Italian and Yugoslavian borders. The leaders of the two advance detachments were *Oberst* Dietl and *Oberst* Schörner.

Following the incorporation of the Austrian Army into the German Army, the Army High Command (OKH) formed two more mountain divisions. The following table depicts the organization and command lists of the three mountain divisions:

1. Gebirgsdivision (_Wehrkreis VII_ Munich)

Div. H.Q.	(Garmisch-P.)	
C.O.:		*Generalmajor* Kübler
Ia: General Staff		*Major* Pemsel

Geb.Jäg.R. 98	(Mittenwald)	*Oberstleutnant* Schörner
Geb.Jäg.R. 99	(Füssen)	*Oberstleutnant* Kress
Geb.Jäg.R. 100	(B. Reichenhall)	*Oberstleutnant* Lanz
Geb.A.R. 79	(Garmisch)	*Oberst* Wintergast
Geb.Pz.Abw. 44	(Murnau)	*Major* Lang
Geb.Pi.Btl. 54	(Mittenwald)	*Oberstleutnant* Zimmer
Geb.NA. 54	(Oberammergau)	*Oberstleutnant* Kleinschroth
Geb.San.Abt. 41	(Garmisch)	*Ob.Arzt* Dr. Bingler

***2. Gebirgsdivision* (*Wehrkreis XVIII* Salzburg)**

Div. H.Q.	(Innsbruck)	
C.O.:		*Generalmajor* Feuerstein
Ia: General Staff		*Major* Degen
Geb.Jäg.R. 136	(Innsbruck)	*Oberst* Eglseer
Geb.Jäg.R. 137	(Lienz)	*Oberst* Schlemmer
Geb.A.R. 111	(Hall)	*Oberstleutnant* Kammel
Geb.Pz.Abw. 47	(Hall)	*Hauptmann* Haidlen
Geb.Pi.Btl. 82	(Schwaz)	*Hauptmann* Drück
Geb.NA. 67	(Innsbruck)	*Hauptmann* Fuhrmann
Geb.San.Abt. 42	(Innsbruck)	*Ob.Arzt* Dr. Lipf

***3. Gebirgsdivision* (*Wehrkreis XVIII* Salzburg)**

Div. H.Q.	(Graz)	
C.O.:		*Generalmajor* Dietl
Ia:		*Oberstleutnant* Ringel
Geb.Jäg.R. 138	(Loeben)	*Oberstleutnant* Weiß
Geb.Jäg.R. 139	(Klagenfurt)	*Oberst* Windisch
Geb.A.R. 112	(Graz)	*Oberstleutnant* Meindl
Geb.Pz.Abw. 48	(Graz)	*Major* Allmendinger
Geb.Pi.Btl. 83	(Graz)	*Major* Klatt
Geb.NA. 68	(Graz)	*Oberstleutnant* Mueller
Geb.San.Abt. 43	(Graz)	*Ob.Arzt* Hinteregger

The necessary supply and motor transport units were not formed until the start of the Second World War.

The personnel strengths of the three divisions were:

	1. Geb.D.	2. and 3. Geb.D.
Officers	640	each 459
Officials	91	each 85
Non-commissioned officers	3,032	each 2,128
Enlisted men	21,193	each 14,516

The personnel strength of all three divisions was 1,558 officers, 261 officials, 7,288 non-commissioned officers and 50,225 enlisted men.

Some of the armament fielded by the three divisions included (not all available weapons are included) 28 heavy field howitzers, 68 mountain guns, 42 light mountain infantry guns, 126 medium mortars, 213 light mortars. The three divisions also had a total of 8,500 pack animals and 4,340 draft horses.

When the 'secret mobilization" began on 26 August 1939, additional units were formed for the three divisions: four mountain horse-drawn columns, 12 small motor transport columns, 7 light motorized infantry columns, 3 small fuel columns, 3 motorized workshop companies and 3 supply companies. Then the Second World War began.

The Second World War

After completing their mobilization preparations, on 27 August 1939 the three German mountain divisions were transported through the protectorate of Bohemia-Moravia into Slovakia. The 3. Geb.D. was the first unit to reach its assembly area just west of the Hohen Tatra. On the day before the war began the 2. Geb.D. was aboard trains bound for Preszow, while the 1. Geb.D. was still far behind between Vienna and the Jablunka Pass.

The three divisions were part of the command of the *14. Armee* which was to drive into southern Poland. Their assignment was to cross the border on the army's right wing between the Jablunka Pass on the left and the Dukla Pass on the right. Stated objectives were the crossing of the Carpathians and the capture of Przemsyl and Lvov.

The 3. Geb.D. crossed the border on the first day of the war and took Neumarkt, whereas the other two divisions were still en route. Eight days later the three divisions were advancing southwest of Przemsyl in line abreast (from right to left: 1., 2. and 3. Geb.D.). Polish resistance was very strong in the Sanok area. There the 3. Geb.D. was halted and turned to the south, while the other two divisions continued to advance east. While the 5. Pz.D. and two infantry divisions covered the *Armee*'s right flank, the 1. and 2. Geb.D. stormed toward Lvov. After overcoming fierce opposition, battle groups of the 1. Geb.D. entered Lvov.

In keeping with the terms of the German-Russian treaty, the mountain troops had to withdraw from the city, which was then occupied by the Red Army. On 27 October 1939 *Generalmajor* Kübler, the commander of the 1. Geb.D., became the first member of the *Gebirgstruppe* to receive the *Ritterkreuz* (Knights Cross of the Iron Cross).

After the Polish Campaign the warring states stopped seeking a political solution to the hostilities. The German Reich began forming new army, navy and air force units. Home industry was converted entirely to war production. The Allied diplomats looked for allies and found them in Germany's neighbor states.

Germany initiated military preparations to defeat France – its only continental foe at that time – however they were bedeviled by the onset of bad weather in autumn 1939, the indecision of the German Supreme Commander, and the political determination of the British. Preparations for the German offensive against France were therefore postponed until spring 1940.

After a meeting of the Allied war council in Paris on 5 February 1940, plans were put into effect to occupy the north of Norway. British warships patrolled the northern Baltic and one of their victims was the German transport vessel *Altmark*.

German preparations for appropriate countermeasures began on 21 February and Hitler signed the necessary operations plans on 1 March 1940. On 2 April he issued orders for "Operation *Weserübung*", the occupation of Denmark and Norway, to commence on 9 April.

The 3. Geb.D. left its barracks in March and transferred into the Nauen-Döberitz-Fronau-Werneuchen-Bernau area. The heavy weapons and materiel were loaded aboard trains and brought to transport ships. Several units remained behind and were transferred to other army units including the 2. Geb.D., which had also been transported to northern Germany.

On 6 April the *Gebirgsjäger* units were brought to Bremen. Headquarters 3. Geb.D. and Geb.Jäg.R. 139 boarded ships of the 1st Destroyer Flotilla (*Kapitän zur See* Bonte) and the troops of the 2. Geb.D. the cruiser *Blücher* and other vessels of the *Kriegsmarine*. The ships left port in Force 9 winds and headed into the rough northern Baltic. The storm lasted several days and resulted in the first casualties – 10 *Jäger* were washed overboard.

On 9 April 1940 troops of the 2. Geb.D. (including Geb.Jäg.R. 138 of the 3. Geb.D.) went ashore in Oslo. The main body of the 3. Geb.D. was unloaded in Narvik. This marked the beginning of the bitter battle for Narvik and the so-called "ore railway" in the most northern theater of operations in Europe. The *Jäger* had to wage a bitter, bloody battle in ice-cold winter weather against a foe (British, French, Poles) superior in numbers and materiel.

It was not until one month later, by which time the 2. Geb.D. had advanced to Trondheim, that an attempt was begun to relieve the troops surrounded by the Allies at Narvik. At the end of May the enemy troops, altogether 28,000 men, evacuated the city and gave up the battle as lost.

The campaign was over. The German mountain troops had completed their mission in spite of unimaginable difficulties, harsh mountain terrain, supply problems and heavy casualties. Seven officers from the two divisions were awarded the Knight's Cross of the Iron Cross.

Based on the lessons learned in the campaign, the OKH decided to reorganize the mountain troops in Norway into a mountain corps (*Gebirgskorps*). In Germany a new *Gebirgsdivision* with the number 5 was to be formed from the elements of the 3. Geb.D. which had been left in the homeland. The formation process was halted in June, however, since the military situation had changed fundamentally with the invasion of Western Europe. The newly-formed *Gebirgskorps* under *Generalleutnant* Dietl assumed command of the remaining German troops in northern Norway. A new *Gebirgsdivision* with the number 6 had meanwhile been created in the Reich and it saw action in the Vosges in the final days of fighting in the west.

With the battle in Norway still raging, on 10 May 1940 the *Wehrmacht* launched its campaign against the Netherlands, Belgium, France and the British Expeditionary Force. The only element of the *Gebirgstruppe* to take part in the campaign was the 1. Geb.D.

Under the command of *Generalleutnant* Kübler, when the campaign began the division was part of the *XVIII. Armeekorps* of the *12. Armee*, whose assignment was to guard the flanks of the panzer units advancing through northern France to the Channel. On 14 May the 1. Geb.D. advanced to the Meuse near Fumay south of Sedan. Three days later it reached the Oise-Aisne Canal and together with the *Maschinengewehr-Bataillon 2* held a 35-kilometer front.

In the second phase of the campaign, on 5 June the division crossed the Oise-Aisne Canal, suffering heavy casualties. It was now under the XXXXIV. AK of the *6. Armee*. The *Gebirgsjäger* broke through the positions at the Chemin des Dames, the Marne and the Seine, and advanced as far as Gien on the Loire. Three of the division's officers won the Knight's Cross in this campaign.

Under the command of *Oberst* Schörner, the 6. Geb.D., which had been formed during the campaign, was committed during the last days of fighting. It was initially held in reserve by the *7. Armee*, crossed the Upper Rhine and in the final days of the campaign advanced into the Vosges.

Since there was no prospect of peace after the French Campaign, the *Wehrmacht* was enlarged enormously and the war economy intensified. The new formations created by the army also included a strengthening of the *Gebirgstruppe*, whose next task was to be the invasion and capture of Gibraltar. This operation was called off, however, as the German command set its sights on the Soviet Union. Subsequently two more mountain divisions were formed.

The 4. and 5. Geb.D. were established in the summer of 1940, mainly at the Heuberg training grounds. The necessary units were taken from the infantry divisions, which were reformed either as panzer or light divisions.

After the necessary training and unit exercises, in March 1941 the two divisions were transferred to Bulgaria as a so-called "Instructional Force". The Balkan Campaign began on 6 April 1941 with the 4., 5. and 6. Geb.D. taking part in the war against Yugoslavia and Greece.

While the 4. Geb.D. engaged the Yugoslav Army, the 6. Geb.D. advanced against the Metaxas Line in northern Greece and the 5. Geb.D. took part in the conquest of Crete.

The struggle for the Mediterranean island went into the history books as an example of cooperation between mountain troops and parachute troops. In spite of heavy losses, elements of the 5. Geb.D. under the command of *Generalmajor* Ringel, who assumed command of all the troops which landed in the western part of the island, stormed the capital city and forced the recently-landed New Zealand troops back into the sea.

While the sound of fighting was still ringing in the Mediterranean, the *Wehrmacht* was preparing for the campaign against the Soviet Union. With the exception of the 5. Geb.D. on Crete, the divisions of the *Gebirgstruppe* were ready to march.

The 1. and 4. Geb.D. were part of Army Group South, whose *17. Armee* had taken up position in southeast Poland and Slovakia. The mountain divisions under its command were part of the newly-formed *XXXXIX. Gebirgskorps*, which occupied jump-off positions east of the San on the Russian-German border. The 1. Geb.D. was in the front line with the 4. Geb.D. held in reserve.

At the same time *Gebirgskorps Norwegen* (2. and 3. Geb.D.) was in the northernmost part of Norway, right on the Arctic Ocean. The mountain corps mission: advance along the coast of the Arctic Ocean and take the important port of Murmansk.

When the Eastern Campaign began early on the morning of 22 June 1941, the 1. Geb.D. attacked toward Lvov for the second time since 1939. The battalions of the 4. Geb.D. followed two days later and on 28 June the two divisions began the assault on the Ukrainian city. While the 4. Geb.D. advanced frontally against Lvov from the west, the *Jäger* of the 1. Geb.D. rushed into the city from the northwest and on 30 June raised the Reich war flag on the central building of the Soviets.

The *Wehrmacht* (OKW) communiqué announced the attack in the Polar region:

> "At 24:00 hours on 29 June German mountain troops attacked in the direction of Murmansk in the glow of the midnight sun over tundra still covered in snow."

After initial success, the German-Finnish offensive bogged down along the Russian-Finnish border. After tough, changeable fighting, the advance by the *Gebirgskorps* under the command of *General der Gebirgstruppen* Dietl ended in a bridgehead east of the Liza. The front on the Arctic Ocean front froze into positional warfare. In September the battered 3. Geb.D. was relieved by the 6. Geb.D., which had been transported from the Reich. Only the Geb.R. 139, which one year earlier had occupied and defended Narvik, remained in the north.

In the following winter the *Oberkommando der Wehrmacht* (OKW = Armed Services High Command) reorganized the land forces in Norway. The former *Gebirgskorps Norwegen* was renamed *Armeeoberkommando Lappland* and on 22 June 1942 was reorganized as Headquarters of the *20. Gebirgsarmee. Generaloberst* Dietl remained as commander-in-chief.

The other two divisions of the *Gebirgstruppe*, the 1. and 4., remained under the command of the *17. Armee*. They took part in the rapid advance through the Ukraine in 1941. Under a blazing sun the *Gebirgsjäger* advanced over dust-shrouded roads and fields, past Vinnitsa and Kirovograd, took part in the battle of encirclement at Uman, crossed the Dniepr, reached the Sea of Azov and fought their way through mud and snow to the Mius. Then came the horrible Russian winter and the eastern army's advance was halted.

When the winter storms ended in early 1942 and spring spread over the land, the eastern army launched another mighty offensive in the Ukraine. The following is an extract from the order for Army Group A, which included the *XXXXIX. Geb.K.* (*General der Gebirgstruppen* Konrad) and the 1. and 4. Geb.D.:

> "...take possession of the entire east coast of the Black Sea...with another group of forces...advance through the western Caucasus and take the Black Sea coast..."

The area assigned to the German forces was part of Asia. Broad steppes, numerous rivers, few sources of drinking water, blazing heat to 40 degrees, sandstorms, steppe fires, downpours and endless dust – these were the adverse conditions through which infantrymen, tank crews and mountain troops now had to advance. Then the snow-covered peaks of the Caucasus reared up before them.

Advancing on the left wing of the *17. Armee*, the XXXXIX. Geb.K. under *General der Gebirgstruppen* Konrad (two infantry divisions plus the 1. and 4. Geb.D.) crossed the several-thousand-meters-high mountains. On 17 August a battle group of the 1. Geb.D. stormed the "Elbrus House" at an altitude of 4 200 meters. Just four days later a mixed battle group of the 1. and 4. Geb.D. stood atop Elbrus (5 633 meters) and raised the Reich war flag.

The southwards advance continued, but just 28 kilometers from the objective of Suchum the attack was halted by fierce enemy resistance and lack of supply. The two mountain divisions went over to the defensive on either side of Elbrus.

At that time the seven mountain divisions were active in every sector of the Eastern Front. The 1. Geb.D. was on the defensive in the Caucasus; the 2. Geb.D. was fighting in the tundra of northern Finland; the 3. Geb.D. was en route from Norway to the Eastern Front, where it arrived in October; the 4. Geb.D. was defending itself against the attacking Soviets in the Caucasus; having been brought back up to strength; the 5. Geb.D. had been transported to the Leningrad front; the 6. Geb.D. was on the extreme left flank on the coast of the Arctic Ocean, where it had been since November 1941. The 7. Geb.D., formed in November 1941 by reorganizing the *99. leichte Division*, had been at the front in central Finland since February 1942.

1943 brought a turning of the tide on the Eastern Front. The Red Army launched its major offensive in the southern sector of the front. This huge operation was interrupted briefly by "Operation Zitadelle". The German attempt to encircle and destroy the Soviet forces in the Kursk salient, in which the mountain troops played no part, proved a failure.

The withdrawal from the Caucasus began on 1 January 1943, and four days later both *Gebirgsdivisionen* left their positions there. For the *Jäger* the weeks and months that followed were filled with tremendous hardships. Not only did the troops have to remove heavy weapons which had been emplaced for months, they also had to transport them out on pack animals over ice- and later mud-covered mountain passes. Once these passes, all of which were at elevations of at least 4 500 meters, had been reached, sleighs and later prime movers offered better means of transport. Finally, thanks to the skill of the commanders and the determination of the officers and men, the retreat was completed and hardly a single gun fell into enemy hands.

In spite of the Red Army's superiority in the field, the German troops were able to withdraw in a relatively orderly fashion. The right wing of the eastern army pulled back to the so-called "Kuban Bridgehead", where the 1. and 4. Geb.D. took up defensive positions in swampy terrain. While the 4. Geb.D. remained in the southern sector of the Eastern Front until the end of the war, in March-April 1943 the 1. Geb.D. was transferred to the Balkans to join the security forces there.

In 1943 the remaining units of the *Gebirgstruppe* remained in their sectors of the front in the Soviet Union and Finland. The only exception was the 5. Geb.D., which left its positions near Leningrad in the winter of 1943-44 and was transported to Italy.

A new theater had been created in Italy after the Allies landed in Sicily and subsequently crossed over to the mainland. By spring 1944 the German Army had been forced to give up southern Italy and was striving to hold a line from Monte Cassino to the Adriatic coast.

The 5. Geb.D. was the only mountain unit in Italy, initially holding positions in the Gari sector near Monte Cassino. It was subsequently moved to the Adria sector and in September 1944 reached the area allocated to it between Small St. Bernhard and Large St. Bernhard and Mont Cenis. The average elevation of this mountainous region on the Italian-French border was 3 500 to 4 500 meters. The 5. Geb.D. remained there to the end, and in the final months of the war it was joined by the 157. Reserve.D. Formed in late autumn 1942, in February 1945 the 157. Reserve.D. was renamed the 8. Geb.D.

In the last two years of the war there were only two theaters for which the German mountain troops had been equipped. One was the Finnish-Russian theater, where three *Gebirgskorps* (XVIII., XIX., XXXVI.), three army *Gebirgsdivisionen* (2., 6. and 7.) and one *Waffen-SS Gebirgsdivision* (7.) saw action from 1943 to 1945.

The other theater encompassed the Carpathians, with towering snow-capped peaks, deep, narrow mountain passes and wretched villages, and the mountains and valleys of the Balkans. While the 1. Geb.D. fought for the Carpathian passes, since 1943 the 1. Geb.D. had been battling Tito's partisan army in Serbia, Albania and Montenegro.

It was in the Balkan theater that the last German mountain division with the high "House Number" 188 saw action. Formed as a reserve division in Italy in October 1943, it was initially under the command of the Commander Adriatic Coastal Region and in August 1944 was transferred to Istria. In February 1945 the division was reorganized as the 188. Geb.D. and it participated in the struggle against the partisans until the end of the war.

After Finland surrendered and its army took up arms in support of the Red Army, the German Army was forced to leave Finland and withdraw to Norway. The 2. Geb.D., which was stationed in the extreme north, left the country at the end of 1944 and was rushed to the Alsace-Lorraine region. This mountain division was the only one of its kind to fight against the western Allies.

The OKH's final order of battle dated 12 April 1945 lists the following areas of operations for the mountain divisions:

Eastern Front:	Army Group South: 1. Geb.D.
	Army Group Center: 3. and 4. Geb.D.
Norway:	20. Gebirgsarmee: 6. and 7. Geb.D.
	and Geb.Brigade 139
Balkans:	Army Group E: 188. Geb.D.
Italy:	Army Group C: 5. and 8. Geb.D.
Western Front:	Army Group G: 2. Geb.D.

Senior Command Units

Armeeoberkommando: 20. Gebirgsarmee (Headquarters, 20th Mountain Army)

In January 1942 the command unit of the *Geb.K. "Norwegen"* was expanded into the *Armeeoberkommando (AOK) "Lappland"* (sometimes also called *AOK "Norwegen"* or *Stab "Finnland"*). On 20 June 1942 it was renamed *AOK 20. Gebirgsarmee*. The army headquarters was responsible for the Finland-Norway theater.

Commander-in-Chief (OB):

Generaloberst Dietl; (from July 1944) *Generaloberst* Rendulic; (from January 1945) *General d. Geb.Truppen* Böhme.

Chief of Staff:

Generalleutnant Jodl, F., (from March 1944) *Generalleutnant* Hölter.

Generalkommandos der Gebirgskorps (Geb.K.) (Headquarters of Mountain Corps)

No.	Formed	Commanding General
XV.	1943	Gen.d.Inf. Lüters, Gen.d.Inf. von Leyser, Gen.d.Pz.tr. Fehn.

XVIII.	1941	Gen.d.Inf. Böhme, Gen.d.Geb.Tr. Eglseer, Gen.d.Inf. Hochbaum
XIX.	1941	Gen.d.Geb.Tr. Schörner, Gen.d.Geb.Tr. Jodl.
XXI.	1941	Gen.d.Art. Bader, Gen.d.Pz.Tr. Fehn, Gen.d.Inf. von Leyser, Glt. Ludwiger
XXII.	1943	Gen.d.Geb.Tr. Lanz
XXXVI.	1941	Gen.d.Inf. Weisenberger, Gen.d.Geb.Tr. Vogel
XXXXIX.	1941	Gen.d.Geb.Tr. Kübler, Gen.d.Geb.Tr. Konrad, Gen.d.Art. Hartmann, Gen.d.Geb.Tr. von le Suire
LI.	1943	Gen.d.Geb.Tr. Feurstein

The battalion- and company-strength units under the command of the general staff of a corps headquarters came from various arms, how ever with the exception of the artillery commander they bore the same number.

Corps	Corps Units	Artillery Commander	Military District Responsible for Replacements
XV.	415	415	XVIII.
XVIII.	418/449	418/449	XVIII.
XIX.	419	463	XVIII.
XXI.	421	421	XIII.
XXII.	422	422	VII.
XXXVI.	436	109	VIII.
XXXXIX.	449	132	VII and Bohemia/ Moravia
LI.	451	451	VXII.

The Divisions

1. Gebirgsjägerdivision

The history of the *Gebirgsjägertruppe* began on 12 October 1935 when the *Gebirgsjägerregiment* (*Geb.Jäg.R.*) *98* was formed. The regimental headquarters, *III. Bataillon* (Btl.) and the anti-tank company moved into the recently-constructed barracks in Mittenwald. The I. Btl. was stationed in Garmisch-Partenkirchen, II. Btl. in Lenggries and the Replacement Battalion in Memmingen.

By 9 April 1938 the remaining units of the new *Gebirgstruppe* (Geb.Jäg.R. 99, 100, *Geb.-Artillerie-Regiment (Geb.AR.) 79, Panzer-abwehr-Abteilung (Pz.Abw.Abt.) 44, Gebirgs-Pionierbataillon (Geb.Pi. Btl.) 54* and the other divisional units with the number 54 formed the first mountain division.

The division headquarters under the command of *Generalmajor* Kübler was stationed in Garmisch-Partenkirchen. Geb.Jäg.R. 99 was based in Füssen, Geb.Jäg.R. 100 in Bad Reichenhall, Geb.AR 79 in Garmisch, Geb.Pz.Abw.Abt. 44 in Murnau, Geb.Pi.Btl. 54 in Oberammergau and the *Gebirgs-Sanitätsabteilung 41* (Mountain Medical Battalion) also in Garmisch.

A report on the division from those days declared:

> "The 1. Geb.D., an elite unit of soldiers who are also enthusiastic mountain climbers. The Edelweiss on their mountain caps symbolizes courage, earthy humor, affinity with nature, comradeship and raw sincerity."

The 1. Geb.D. saw its first action prior to its actual formation. Early on 6 March 1938, at the start of the *Wehrmacht*'s entry into Austria, three motorized battle groups from the division crossed the border and drove through the country to the Italian, Yugoslavian and Hungarian borders. The three regimental commanders (*Obersten* Schörner, Dietl and Konrad) established contact with their respective border outposts. After a few days and weeks as occupation troops the units returned to their home garrisons.

"Peacetime" was not to last much longer. With the start of the "secret mobilization" on 26 August 1939 the units were again brought to march readiness. Beginning on that day the 1. Geb.D. was transported to southeast Poland and Slovakia. The Second World War began at dawn on 1 September 1939.

The division advanced out of the Zborow-Lubotin area (Slovakia) in the direction of Krempa, fought its way across the Dukla Heights into the Rymanow area, forced a crossing of the San near Sanok, advanced rapidly on Lvov and took the city. There the *Gebirgsjäger* were forced to halt, for battle groups of the Red Army were advancing from the east. According to the terms of the German-Russian treaty eastern Poland belonged to the Soviets.

Division commander *Generalleutnant* Kübler became the first member of the *Gebirgstruppe* to be decorated with the Knight's Cross of the Iron Cross.

After the Polish Campaign the division was transferred across the Reich into the Eifel region. It left behind 484 fallen on the Polish battlefield; as well, 918 soldiers had been wounded in the campaign.

In October 1939 the 1. Geb.D. was placed under the command of the *4. Armee* as reserve. It was then attached to Headquarters, *12. Armee* (AOK 12) in the same area of operations from December of that year to spring 1940.

When the Western Campaign began on 10 May 1940, battle groups of the division advanced across the Luxembourg border near Dahnen and four days later reached the Meuse south of Sedan. The *Gebirgsjäger* crossed the river near Fumay and advanced toward Hirson. From there, after taking Mondrepuis, the division advanced to the Oise-Aisne Canal near Coucy-le-Chateau. As part of the *XVIII. Armeekorps* it helped storm the Chemin des Dames. The second phase of the campaign was launched from there. In fierce fighting which cost the division 500 killed and wounded, the division crossed the Oise-Aisne Canal and the Aisne near Soissons. In the follow-up advance it forced the Marne near Chateau-Thierry. Six days after the start of the second phase of the campaign, the Seine was crossed northwest of Romilly, followed by a rapid advance to the Loire. When the halt order was issued, the division's advance guard under *Oberstleutnant* Lang was eight kilometers from Gien.

In the course of the campaign the *Gebirgsjäger* had covered a total distance of 1 826 kilometers, most of it on foot. Left along the way were 446 fallen comrades, while another 1,308 *Gebirgsjäger* were wounded.

After the campaign the division was one of the first major units of the army to return home. The Geb.Jäg.R. 100 and I./Geb.AR. 79 left the division and were assigned to the newly-formed 5. Geb.D. From this point on the 1. Geb.D. consisted of just two regiments, like all of the army's other mountain divisions.

The 1. Geb.D. returned to France at the end of the year. The various units arrived in the Besancon area to take part in an attack on Gibraltar. Preparations for the attack were supposed to be completed by January 1941. In addition to the division, the *Wehrmacht* High Command (OKW) massed heavy artillery, pioneer and air units for the attack, however "Operation Felix", as the attack on the British fortress was code-named, was called off. The Spanish government was not prepared to allow German troops to pass through the country.

After a few weeks the *Gebirgsjäger* returned to the Reich and were immediately transported to Austria. There the division was placed under the command of the XXXXIX. Geb.K. of the *2. Armee*.

The army assembled its attack units for the assault on Yugoslavia. The Balkan Campaign began on 6 April 1941 with heavy air raids on Belgrade. The 1. Geb.D. attacked through Cilli, Karlstadt, Agram and Bihac. The enemy offered fierce resistance but was driven back everywhere. The division was halted after reaching Bihac and in May was transferred into Slovakia, where the *Gebirgsjäger* had been stationed prior to the attack on Poland.

The corps and divisions of the *17. Armee* had been massing in Slovakia since the beginning of the year. Included in these was the XXXXIX. Geb.K. with the 1. and 4. Geb.D.

The German attack on the Soviet Union began on 22 June 1941. In fierce fighting the *Gebirgsjäger* overcame the Soviet border fortifications and for the second time advanced on Lvov. The city was taken, but at a cost. The advance continued toward the "Stalin Line", which was breached near Bar. The next objective was Vinnitsa. Enemy resistance there was much stronger and after heavy rainfalls the German troops encountered a new foe: the Russian mud. The motorized columns became bogged down. The mud clung to the mountain boots and trousers of the *Gebirgsjäger* and only the brave mules and horses were able to make any headway.

The Soviet position on the Bug near Brazlaw was finally overcome and with the return of fine weather the enemy was pursued through Gaissin into the Uman area. The 1. and 4. Geb.D. helped close the pocket around the enemy troops. The *Gebirgsjäger* then turned to the east and southeast and advanced on the Dniepr.

After taking Berislav the regiments and battalions advanced into the Nogai Steppe. There were offensive and defensive battles between

Timoshevka, Balki and Verkhne Tokmak. By this time autumn had arrived; the nights were perceptibly colder and the rain and mud returned. In October the advance resumed in the direction of the Mius. The river was crossed between Diakovo and Bobrikovo. But by then the German forces in the south of the Eastern Front, like those in the center and the north, were exhausted.

In December 1941 the 1. Geb.D. was forced to withdraw to the west bank of the Mius. The opposing Soviet troops tried almost daily to drive the *Gebirgsjäger* from their positions. Then in mid-February 1942 orders were suddenly received to move to the Samara. There the troops dug in again, and positional warfare began in the Barvenkovo-Petropalovka area.

In the spring the Red Army launched a skillful offensive and broke into the German positions east of Kharkov. The 1. Geb.D. was withdrawn from its previous main line of resistance and attached to the *III. Panzerkorps* (Pz.K.). In May the corps attacked the Russian divisions which had broken into the German front and restored the main line of resistance. The *Gebirgsjäger* saw action on the Bereka. After the battle the fighting settled down into positional warfare in the Balakleya area.

At the end of July the *Gebirgsjäger* were finally able to go on the offensive again. In the south the Eastern Army attacked in the direction of the Volga and the Caucasus. The division was now part of the XI. AK. and forced the crossing of the Donets near Kamensk. The division was subsequently subordinated directly to the *1. Panzerarmee*, which had assembled itself in the Rostov area on the lower Don in preparation for a drive into the Caucasus.

The 1. Geb.D., now once again under the XXXXIX. Geb.K., attacked from the Donets plain toward the Kuban. It crossed the huge river and swamp region and drove into the steppes of the Caucasian foothills. Then the terrain became rockier and more mountainous and the snow-covered peaks of the Caucasus Mountains became visible in the distance. The *Jäger* were now fighting in familiar mountain terrain. The enemy troops retreated everywhere. The mountain troops stormed the Elbrus House, capturing the Russian garrison, and on 21 August 1942 a battle group from the division and a squad from the 4. Geb.D. raised the Reich War Flag and the banners of both divisions on the 5 336-meter Mount Elbrus.

The soldiers of the 1. Geb.D. were standing on the "Throne of the Gods", the name given to the High Caucasus by the inhabitants of the mountainous region.

The advance was now over, however. The supply system had broken down because of the huge distances involved and Soviet resistance grew stronger from week to week.

Before the end of September the 1. Geb.D. had to evacuate the Elbrus region and was transferred into the Maykop area. The German forces were reorganized for a final drive to the Black Sea coast near Tuapse, however the superiority of the Soviet defenders was such that the mountain troops were unable to gain any further ground. In November they had to abandon their positions and from that point on there developed fierce fighting withdrawals which lasted until the end of the year.

At this time the division commander, *Generalmajor* Lanz, left the division and transferred command to *Oberst* Stettner Baron von Grabenhofen, who had been commander of the I./Geb.Jäg.R. 98 at the start of the war.

The new year of 1943 saw the 1. Geb.D. forced to withdraw from its mountain positions in the ice and snow. Casualties were heavy. Beginning on 1 January 1943 the German troops on the left wing withdrew from the mountain and steppe regions of the Caucasus.

The individual battalions withdrew into the Slavyansk area with the pursuing enemy hard on their heels. To date the *Jäger* had covered 4 995 kilometers since the start of the eastern campaign. The division was now "burnt out" and beginning in March it was withdrawn from the front. The units assembled in the rear and were transported by rail to the Balkans.

The month of April 1943 found the units of the division in the Nisch area. Men, materiel and weapons were delivered there, and in the weeks that followed the division was turned back into a combat unit once again. The I./Geb.Jäg.R. 98 was taken from the division and placed under the direct command of the OKH as *Hochgebirgs-Jäg.Btl. 43* (High-Mountain Infantry Battalion 43). As a replacement unit the division received the Geb.Jäg.Btl. 54. The battalion had previously seen action as the independent *Hochgebirgs-Jäg.Btl. 2*.

The Balkans had already become a theater of war by this time, with Tito's partisan movement receiving heavy weapons, rations and war materiel from the Soviet Union and Great Britain. The battle groups of the 1. Geb.D. had to seek out and engage partisan groups in the mountains of Montenegro, often completely on their own.

The 1. Geb.D. had left the eastern theater for good and from this point on it fought virtually as an independent unit in the "gorges of the Balkans." At the end of June – beginning of July 1943 the division was suddenly transferred to central and southern Greece. There the battle-tested *Gebirgsjäger* were employed in the coastal defense role from July to October.

As a result of the sudden change of sides by Germany's Italian ally, the units of the division were abruptly forced to take Italian troops prisoner, seizing their weapons and other war materiel.

In November 1943 the 1. Geb.D. was transported to northern Greece and attached to the XXII. Geb.K., whose units were in action against partisan bands in Bosnia. The fighting went on for months and the division's battle groups pushed through Bosnia to Croatia. It was months of warfare against an almost invisible foe, a cruel enemy who offered no pardon.

Without warning the order came to enter and occupy Hungary. The German command had learned that the Hungarian government and military were negotiating with the Red Army, and preparations were made for "Operation Margarethe". Units of the army subsequently attacked Hungary from four sides and marched into the country.

Designated Battle Group "A", the 1. Geb.D., reinforced by a motorized bridging battalion, entered Hungary from the southwest from the Virovitica area. The occupation of the country proceeded without a hitch. There was no rest for the *Jäger*, however, for the Red Army was nearing the Hungarian border from the east. The division hastily occupied the Uzsok-Gyimes Pass in order to defend this bulwark against the enemy.

When the situation in this area stabilized, the division was hastily ordered to Montenegro for use against partisans.

The division advanced in battle groups from Macedonia into the border region between Albania and Greece. Its advance was supposed to cut off the partisan units, which were supported by the British, from the coast. Losses were heavy, however the 1. Geb.D. accomplished its role in "Operation Gemsbock". There was little time to settle into the new area, however, for the division was soon moved into the area around Skolpje.

The division found itself engaged in a struggle against a new and vicious enemy. The *Gebirgsjäger* battle groups remained in constant action in the border region of Serbia-Montenegro. In August 1944 the division fought in the area around Nisch, and one month later the first troops of the advancing Red Army were repulsed at the Bulgarian border.

The enemy's superiority was so great that the German troops had to withdraw to the Morawa and then into the Belgrade area. Soviet armored forces succeeded in encircling large elements of the German defenders there, including the division's battle groups. The commander and approximately 5,000 men were trapped in the pocket in and around Belgrade. On 18 October 1944 the German units broke out of the encirclement. Only a few *Gebirgsjäger* managed to reach the German lines (the division commander was not among them).

The remnants of the 1. Geb.D. fought their way back to the Drina and held positions there until the end of February 1945 against attacks by the Soviets and Tito partisans. On 25 February 1945, the division was renamed the *1. Volksgrenadier-Division.* (It was the only division of the *Gebirgstruppe* to receive such a designation.)

The *Gebirgsjäger* took part in the German offensive toward Budapest in spring 1945. After the failure of the last German offensive on the Eastern Front, the division was forced to withdraw toward the German-Hungarian border under the command of the XXIII. Geb.K. of the *2. Panzerarmee.* The seriously weakened unit reached the area around Pöllau then moved as far as the Semmering. On the day of the surrender the last *Gebirgsjäger* laid down their weapons in the Liezen area.

The Commanders

Generalleutnant Kübler, L; *Generalleutnant* Lanz; *Generalleutnant* Stettner Ritter von Grabenhofen; *Generalleutnant* Kübler, J.; *Generalleutnant* Wittmann.

2. Gebirgsjägerdivision

The division was created following the absorption of the former Austrian Army into the German Army in the Innsbruck area. Bolstered by German officers and non-commissioned officers, the mountain units of the former Austrian Army formed the *2. Gebirgsdivision.* The likewise newly-formed *Wehrkreis XVIII* (Salzburg) was responsible for the establishment of the division.

The 2. Geb.D.'s organization and command list was as follows in the summer of 1939:

Division headquarters in Innsbruck:	*Generalmajor* Feuerstein
Geb.Jäg.R. 136 in Kuftsein:	*Oberst* Eglseer
Geb.Jäg.R. 137 in Lienz:	*Oberst* Schlemmer
Geb.AR. 111 in Hall:	*Oberstleutnant* Kammel
Geb.Pz.Abw.Abt. 47 in Hall:	*Hauptmann* Haidlen
Geb.Pi.Btl. 83 in Schwaz:	*Hauptmann* Drück
Geb.NA. 67 in Innsbruck:	*Hauptmann* Fuhrmann
Geb.San.Abt. 42 in Innsbruck:	*Oberarzt* Dr. Lipf

After a brief period of training and unit exercises, in August 1939 the division was transferred to Slovakia, where it assembled in the Käsmark area in preparation for the attack on Poland. When the Second World War began the *Gebirgsjäger* moved across the German-Polish border and fought their way through Dunajec and Nida toward Neu-Sandez, often against fierce resistance. The division then attacked northeast through Jaslo toward Lvov. There the units halted, and after the entry of Soviet troops into Poland they pulled back behind the demarcation line.

After the Polish Campaign the division was transferred into the St. Goar area, where it held security duty and was brought up to strength. The independent units which had been attached to the unit – II./Geb.Jäg.R. 140 (peacetime base Kufstein) and I./Geb.AR. 113 (peacetime base Spittal) – were renamed II./Geb.Jäg.R. 136 and I./Geb.AR. 111, but remained with the division.

Then, without warning, at the end of March – beginning of April 1940 the division was loaded aboard trains and transported to the North Sea coast. From there transport vessels of the *Kriegsmarine* took the Austrian *Gebirgsjäger*, who were totally unfamiliar with sea travel, to Norway. Once ashore, the combat troops had to secure starting positions in the Trondheim area before they could attack the determined Norwegians. After linking up with the infantry units advancing from Oslo, at the beginning of May the division had to begin preparations for an attack toward distant Narvik. At Narvik the units of the 3. Geb.D., together with the crews of sunken German destroyers and several parachute companies under its command, was in a hopeless situation. The 2. Geb.D. was given the task of relieving the "Dietl Group" and recapturing Narvik.

At the end of May the battle groups of the 2. Geb.D. moved out of the area around Namsos and fought their way north, literally meter by meter, hampered by difficult terrain and stubborn British and Norwegian resistance. Elements of the Geb.Jäg.R. 137 under *Oberstleutnant* Sorko were the first to make contact with the soldiers of Battle Group Dietl.

The division had now reached an area which was to become its "second home" in the years to come. After the end of the Norwegian Campaign the battle groups marched further north and did not halt until they reached the area around Kirkenes. There the *Gebirgsjäger* had to become accustomed to the polar conditions of northern Norway. Nevertheless, they held exercises in near peacetime conditions, although patrols monitored the Soviet border and the border troops stationed there.

The 2. Geb.D. – under the command of the *Geb.K. "Norwegen"* together with the 3. Geb.D. – was now in an area of which the later chief-of-staff of the *20. Gebirgsarmee* wrote:

> "The country and climate molded the troops and their battle style. ... The land front extended 650 km from Kiestinki to the Fischer Peninsula, the sea front 600 km from Petsamo Fiord to Nordkyn. Supplying the corps on the Polar Sea was accomplished mainly by sea along the Norwegian coast, while the fronts on the Arctic Circle were supplied by way of the Baltic. However, in the middle of winter (when temperatures reached –40 degrees) and until late in the spring, ice blocked the sea routes to and from Germany. ..."

On 29 June 1941 the OKW declared in its daily communiqué:

> "At 0000 hours on 29 June 1941, in the glow of the midnight sun, German mountain troops launched an attack across still snow-covered tundra in the direction of Murmansk."

The 2. Geb.D. had been ordered to attack from the Paarkkina area and break through the Soviet border fortifications in the Kuosmaivi area. Afterward the attack was supposed to continue east through tundra and primeval forest to the fortified Camp Titovka and across the river of the same name. In spite of the difficult terrain and fierce Soviet resistance, in the first days of July the *Gebirgsjäger* (since autumn 1940 two of the units under the division's command, the Pz.Jäg.Abt. 47 and Geb.AA. 111, had been replaced by the Pz.Jäg.Abt. 55 and Geb.AA., respectively) advanced to the Liza River. The initial attempt to cross the river failed, however.

By mid-July repeated attempts to cross the river line had resulted in considerable losses in men and materiel. Then came the order to pull back into a small bridgehead west of the line Round Lake, Moon Lake and Liza Fiord.

There the division's attack came to a standstill. Some battle groups even had to be turned around following a Soviet landing in Motovski Bay. This group of forces attacked the German mountain troops from the rear. The division subsequently released companies to the newly-formed *Divisionsgruppe "Petsamo"*, whose units sealed off the coast and the Fischer Peninsula.

Mid-August 1941 saw the beginning of positional warfare in Lappland and on the Polar Sea front. Units of the division remained in the small bridgehead east of the Liza. On 8 September battle groups attacked once again in an effort to enlarge the bridgehead. By 20 September they had gained only a few meters of ground and the attack was halted. The 2. Geb.D. went over to positional warfare.

At the end of October some front-line units were relieved by elements of the 6. Geb.D. which had arrived in Norway in August 1941. Those elements which were pulled out of the line received replacement personnel and equipment just behind the lines; however, for the division as a whole the positional warfare in the Liza bridgehead continued. Defensive battles now alternated with small offensive operations. Other battle groups from the division were assigned to counter enemy commandos landed in the rear area by sea or air.

During the winter of 1941-42 it became possible to transfer entire units of the division to rear area and training camps in Norway and northern Finland for several weeks of rest.

The German command authorities now formed the Geb.AOK 20 (Headquarters, 20th Mountain Army) under the command of *Generaloberst* Dietl, which was responsible for all German and Finnish troops in Lappland.

This move simplified the chain of command in northern Norway, the Finnish high command having declared that it would only accept responsibility for the defense of Finland. The *20. Geb.Armee* assumed responsibility for the three sectors of front: Louhi (XVIII. Geb.K. with two divisions), Kandalakscha (XXXVI. Geb.K. with two divisions) and Murmansk (commanded by the XIX. Geb.K). The XIX. Geb.K. had two mountain divisions under its command on the land front and one infantry division and one division group plus units of the *Kriegsmarine* to defend the sea front.

In spring 1942 the Red Army began a series of attacks against the German front between Louhi and Murmansk. There was defensive fighting everywhere; local penetrations by the Soviets were eliminated.

All of the units of the 2. Geb.D. were in the front and in July these repulsed particularly fierce attacks in the Parkkina area. These often costly battles continued until October 1944, however by this time it was obvious that a shift was taking place in the military and political leadership of Finland. The headquarters of the *20. Geb.Armee* devised plans for the continued prosecution of the war after a possible Finnish withdrawal from hostilities.

On 3 September 1944 the OKW, which was responsible for the Norway-Finland theater, ordered the start of "Operation Birch". The two corps in the middle and south of the *20. Geb.Armee*'s front had to evacuate their positions and withdraw through the Rovanniemi area to northern Norway. Initially, the XIX. Geb.K., with the 2. Geb.D. on the right and the 6. Geb.D. on the left, had to hold the existing main line of resistance to cover the withdrawal by the other two corps. Not until one month later, on 4 October 1944, did the corps receive orders to give up the positions it had held for more than three years.

The withdrawal into northern Norway, "Operation Northern Light", could not be launched as planned, as the Red Army had already advanced to the so-called Polar Sea road between Rovanniemi and Kirkenes. On 7 October 1944 the 2. Geb.D. left its positions in the Salmijärvi area and as per orders withdrew southwest to the Polar Sea road.

The Chief-of-Staff of the 20. Geb.AOK, *Generalleutnant* Hölter, wrote in his book *Armee in der Arktis* (Army in the Arctic):

> "The 2. Geb.D. was seriously battered in the initial Russian onslaught. Unused to large-scale warfare, the brave mountain troops from the Alpine lands never fully recovered in the very difficult days of fighting which followed. The division quickly lost ground to the west. On 8-9 October it was already 10 km west of the Isar camp on the hills astride the Lahn road."

The division withdrew in battle groups southwest into the area around Ivalo and Karasjok. As the division neared Luostari it encountered strong Soviet forces determined to break through to the Polar Sea road at any cost. The division's battle groups were forced away to the north and had to abandon Luostari. But when the enemy pressure abated for several days, the corps was able to pull the 2. Geb.D. out of the front and send it via Tarent to Salmijärvi. The division was once again forced to make an about face when the Soviets were able to advance to the Samijärvi-Kirkenes road.

Following the evacuation of Kirkenes, the campaign in the extreme north of Europe was over for the 2. Geb.D. After the Finnish-Norwegian border had been crossed and some semblance of order restored, the OKW ordered several divisions back for the final battle in the Reich. Among them was the 2. Geb.D.

The division completed an arduous march south from Denmark into the Reich. When they arrived, the battle groups were put aboard trains and transported across northern and western Germany into the Upper Alsace. There, under the command of the *19. Armee*, in January 1945 they took part in the defensive effort against French and American forces.

At the beginning of February 1945 the 2. Geb.D. was attached to the LXXXII. AK. of the *1. Armee*. The corps was on the *Armee*'s right wing between the Rhine and the Moselle. The *Gebirgsjäger* fought armored units of the American 3rd Army south of Trier. The Americans fought their way into the former *Westwall* sector near Saarburg and advanced across the Saar toward Zerf, however a counterattack by the division together with elements of the 11. Panzer-Division drove the enemy back.

The main line of resistance was held until mid-March, although the division suffered very heavy casualties. In the withdrawal to the Rhine, some battle groups and individual *Gebirgsjäger* managed to slip through the American lines, which had long since overtaken the division. Carrying their light weapons (the guns and remaining vehicles had been blown up) the *Jäger* slipped through the encirclement and reached the Rhine near Germersheim.

After the German *1. Armee* had assembled in relatively good order east of the Rhine, it withdrew to Baden and Württemberg. The 2. Geb.D. was now under the command of the XC. AK. on the Armee's left wing. The retreat took the *Jäger* from Germersheim to Bruchsal and Ebersberg, through the Odenwald to Miltenberg am Main. The battle groups subsequently pulled back to the western edge of the Kraichgau hill country.

At the beginning of April the division command received orders to defend Heilbronn. For this task several poorly armed *Volkssturm* units and an infantry replacement unit were placed under the division's command. The senseless battle raged for eight days, then the battle groups evacuated the city and withdrew east, then southeast and finally south. Lacking anti-tank and anti-aircraft weapons, the troops could not offer much resistance. American close-support aircraft repeatedly strafed the individual battle groups. Resistance was sporadic, often involving fanatical *Volkssturm* or Hitler Youth units determined to defend their towns. The retreat took the *Gebirgsjäger* through the Crailsheim area to Ansbach and then to Neuburg on the Danube and Freising.

Then, for the first time in many years of war, the once familiar mountain landscapes came into view. In oppressive heat the remnants of the once vital mountain division marched to Reit im Winkel and Kössen, where they were forced to surrender their weapons. The war was over.

The Commanders

Generalleutnant Feuerstein; *Generalmajor* Schlemmer; *Generalleutnant* Ritter von Hengl; *Generalleutnant* Degen; *Generalleutnant* Utz.

3. Gebirgsjägerdivision

Like its sister division, the 3. Geb.D. was formed from elements of the former Austrian Army. Under the command of Wehrkreis XVIII (Salzburg), complete units of the 5th and 7th Austrian Divisions were combined to form the new 3. *Gebirgsjägerdivision.*

The division's organization and command list was as follows in the summer of 1939:

Division headquarters in Graz:	*Generalmajor* Dietl
Geb.Jäg.R. 138 (Loeben):	*Oberstleutnant* Weiss
Geb.Jäg.R. 139 (Klagenfurt):	*Oberst* Windisch
Geb.AR. 112 (Graz):	*Oberstleutnant* Meindl
Geb.Pz.Abw.Abt. 48 (Graz):	*Major* Dr. Allmendinger
Geb.Pi.Btl. 38 (Graz):	*Major* Klatt

Mittenwald, with the mighty Karwendel Mountain in the background, was the birthplace of the German mountain troops. Here the ceremonial swearing-in of recruits takes place on a parade grounds.

The barrack gate of I./Geb.AR. 79 in Bad Reichenhall.

26 August 1939 was the day of the secret mobilization. Ready to march, the troops of Geb.Jäg.R. 98 stand at Mittenwald Station waiting for transport to the east.

5 September 1939: In the blazing heat, the companies of Geb.Jäg.R. 99 cross the German-Polish border. The Second World War had begun. All three German mountain divisions saw action in the Polish campaign in the southern sector.

After a battle. Command post of the Geb.Jäg.R. 98. Second from the right is Oberst Schörner, the later Generalfeldmarschall.

Generalleutnant Dietl, commander of the 3. Geb.D. in the Narvik combat zone. The later Generaloberst and Commander-in-Chief of the 20. Gebirgsarmee in the far north was a symbol of the German mountain troops. The Generaloberst was killed in a plane crash in Styria in June 1944.

On 9 April 1940 ten German destroyers brought elements of the 3. Geb.D. to Narvik in the north of Norway. Here Gebirgsjäger unload material from a destroyer in the port of Narvik.

Gebirgsjäger on ski patrol in the mountains of northern Norway.

A machine-gun position above the port of Narvik, following the Allied landings and the German withdrawal.

Austrian Gebirgsjäger at a small station on the ore railway, the focal point of German and Allied strategy in Norway.

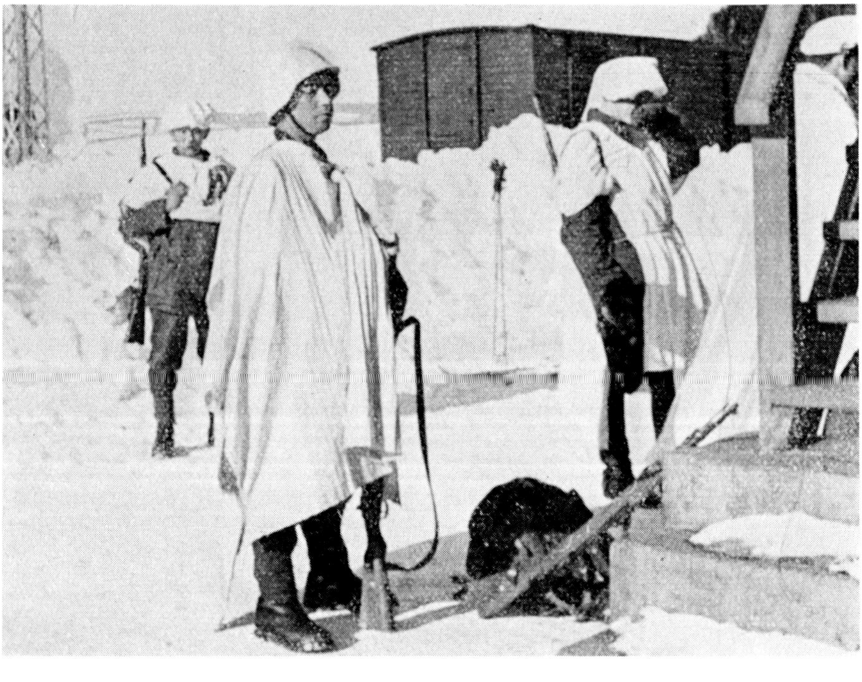

Only the 1. Geb.D. saw action in the western campaign. This photo depicts a mountain battery passing through a French town near Lyon in hot summer weather.

In April 1941 the mountain troops were called into action again for the campaign in the Balkans, this time the newly-formed divisions. Radio station of the 5. Geb.D. at the edge of the Belaschitza Mountains.

Mule transport column moving through the mountains of Greece bound for Athens.

The campaign in the Balkans ended in the early days of May 1941. The German mountain troops had proved themselves again. While the 1. Geb.D. fought the Yugoslavian Army, the 5. and 6. Geb.D. saw action in Greece and on Crete. Here troops of the 6. Geb.D. parade before Feldmarschall List in Athens on 4 May 1941 (to his right and in front is General der Flieger Baron von Richthofen).

At the beginning of the eastern campaign the 1. and 4. Geb.D. took part in Army Group South's advance through Galicia and the Ukraine. The two divisions stormed the city of Lvov at the end of June. This photo was taken in the city; the destroyed GPU House is on the left.

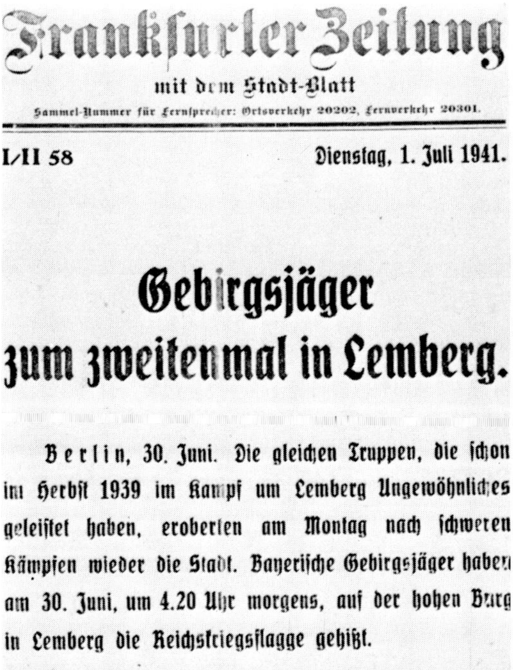

Frankfurter Zeitung

mit dem Stadt-Blatt

Sammel-Nummer für Fernſprecher: Ortsverkehr 20202, Fernverkehr 20301.

I/II 58 Dienstag, 1. Juli 1941.

Gebirgsjäger zum zweitenmal in Lemberg.

Berlin, 30. Juni. Die gleichen Truppen, die ſchon im Herbſt 1939 im Kampf um Lemberg Ungewöhnliches geleiſtet haben, eroberten am Montag nach ſchweren Kämpfen wieder die Stadt. Bayeriſche Gebirgsjäger haben am 30. Juni, um 4.20 Uhr morgens, auf der hohen Burg in Lemberg die Reichskriegsflagge gehißt.

A newspaper report from July 1941 declares that the mountain troops have entered Lvov for the second time.

The German summer offensive of 1942 saw the mountain divisions advance shoulder to shoulder into the Caucasus. On 1 August 1942 a battle group of the 1. Geb.D. stormed the Elbrus House (4 200 meters) and captured the Russian garrison. A few days later a mixed battle group from the 1. and 4. Geb.D. climbed to the top of Elbrus, at 5 633 meters the highest peak in the massive mountain range.

The mountain world of the Caucasus was the mountain troops' battleground in 1942. The planned breakthrough to the coast was frustrated by Soviet superiority in men and materiel as well as by supply problems.

While the 1. and 4. Geb.D. were fighting in the extreme south of the Eastern Front, the 2., 3. and 7. Geb.D. were in constant action in the far north (northern Finland and on the Polar Sea). The photo shows a supply column on its way to the Liza front.

In the far north the troops had to battle nature as well as the enemy. Dense forests, swamps and tundra all presented serious obstacles. Here a corduroy road through the north-Finnish forest.

Gebirgsjäger ski patrol east of Alakurtti.

The Commander-in-Chief of the 20. Geb.Armee, Generaloberst Rendulic, inspects the positions of the 7. Geb.D. (August 1944).

Mountain divisions saw action in northern Italy beginning in November 1943. The photo shows soldiers of the 1. and 4. Geb.D. taking prisoner and disarming Italian troops.

Beginning in 1943 mountain troops saw frequent action in the Balkans in the struggle against the Tito partisans and later the Red Army. Here mountain troops are seen during the retreat through the gorges of the Balkans.

Monument to the fallen of the German mountain troops on the Hohen Brendten near Mittenwald.

Geb.NA. 68 (Graz):	*Oberstleutnant* Mueller
Geb.San.Abt. 43 (Graz):	*Oberarzt* Dr. Hinteregger

After barely fifteen months of training and associated unit exercises, in late summer 1939 the division was transported to southern Poland and Slovakia, detraining in the Rosenberg area. When the Second World War began, advance detachments of *Gebirgsjäger* crossed the border, preparing the way for the following units to advance across the Tatra. The division suffered its first casualties near Dolna and Walka. After the stubborn Polish resistance had been broken, the advance proceeded rapidly across the Dunajec and the Tarnow to the San River near Sanok. Battle groups of the division turned toward Lvov, but then the advance was halted. The Red Army had entered eastern Poland.

In October 1939 transport trains carried the 3. Geb.D. across Germany to the Eifel region. There the division was initially under the command of the AOK 6, later the AOK 16, which committed the unit to security duties in the Traben-Trarbach area.

The various units of the division now received further combat training in anticipation of an attack across the western border. All of these preparations proved pointless, however, for in March 1940 the division unexpectedly received orders to transfer to the greater Berlin area.

The division's troops were temporarily quartered in barracks and other quarters in the area of Nauen, Döberitz, Fronau, Werneuchen and Bernau. The Geb.Jäg.R. 139 had to relinquish all of its heavy weapons and equipment, and together with the division headquarters it boarded trains of the *Reichsbahn* (German State Railway). The division's heavy equipment followed the same way and was subsequently loaded aboard transport vessels of the *Kriegsmarine*. (The majority of these ships were lost at sea together with their cargoes.)

The Geb.Jäg.R. 138 followed with other elements of the division, however it had a different destination, namely Trondheim. The rest of the 3. Geb.D. remained in the Berlin area and was used in the formation of three new mountain divisions, while other units later served as army troops.

The division headquarters and the Geb.Jäg.R. 139 arrived in Bremen on 6 April 1940; from there they proceeded to Wesermünde. Waiting there were ten destroyers of the *Kriegsmarine* under the command of *Kapitän zur See* Bonte. The *Gebirgsjäger* were put aboard the ten cramped combat vessels, with approximately 200 to 250 men with small arms on each destroyer. The division headquarters went aboard the command destroyer.

The destroyers sailed at 0100 hours on 7 April. Two hours later they rendezvoused with the battleships *Scharnhorst* and *Gneisenau* and the heavy cruiser *Hipper*, which were waiting in the southern North Sea. The fleet unit headed into heavy seas and a rising storm with Force Nine gales. The storm produced the first casualties: a total of ten *Gebirgsjäger* were washed overboard, the first victims of the new campaign.

The German naval unit continued north unhindered. Northwest of Trondheim contact was made with British warships. The battleships then turned away to the northwest while the cruiser *Hipper* was detached to Trondheim with the troop carriers.

The ten destroyers continued north and on the morning of 9 April, after destroying two armored coastal vessels, landed in Narvik harbor as per orders. Some of the destroyers turned away to the northeast and put the *Gebirgsjäger* ashore near Elvegaardsmoen. There the mountain troops encountered resistance for the first time, from troops of the 6th Norwegian Division based at Camp Elvegaardsmoen.

Once ashore the *Gebirgsjäger* took possession of Narvik, advanced along the ore rail line and dug in on the Framnes headland between Rombakenfjord and Beisfjord. The companies were still not in established positions when, the next day, British naval units blocked the entrance to the west fjord and sent destroyers in to Narvik. All ten German destroyers were sunk that day. The surviving crews placed themselves under the command of *Generalleutnant* Dietl and from then on assisted in the defensive effort by the "Dietl Group".

The battle for Narvik and the ore rail line began on 14 April 1940 when the British 21st Guards Brigade went ashore following the German evacuation of Narvik. The next day 6,000 troops of the British 49th Division came ashore. First contact with the "Dietl Group" took place on 24 April 30 kilometers northeast of Narvik.

This marked the beginning of a four-week battle for Narvik and the ore rail line to the Swedish border. The "Dietl Group", which also included parachute troops, fought alone against Norwegian, British, French and Polish units, but in spite of losses and hardships it was able to hold the rail line. Hitler even considered ordering Dietl and his soldiers to withdraw to Swedish territory, however the general staff officers in the OKW and the stubborn *Gebirgsjäger* general held firm. When battle groups of the 2. Geb.D. approached from the south, the Allies withdrew from Narvik. The battle was won.

Those elements of the 3. Geb.D. not committed at Narvik (and not attached to other units) were assigned to the 163. and 196. ID. At the start of the Norwegian Campaign these divisions landed in Oslo and Trondheim respectively. From there the units advanced into the area around Trondheim, Dombas and Mo. Finally these battle groups of the 3. Geb.D. joined the forces of the 2. Geb.D. relieving Narvik.

Many members of the division were decorated after the campaign. Their commander was promoted to *General der Gebirgstruppen* and became the first soldier of the *Wehrmacht* to receive the Knight's Cross with Oak Leaves. Six officers of Geb.Jäg.R. 139 were awarded the Knight's Cross.

After the 3. Geb.D. was reunified at the end of May – beginning of June 1940, several changes were made in its organization. The III./Geb.AR. 112 and the Geb.AA. 112 had to be released for the formation of the 6. Geb.D. The elements of the Geb.Jäg.R. 139 which had been left behind in Berlin also joined the new division. In their place, in autumn 1940 the division received the newly-formed *Geb.Radfahr-Btl. 68* (Mountain Bicycle Battalion).

Generalmajor Ringel assumed command of the division, which went on occupation duty in northern Norway until May 1941, when it moved to the Finnish-Russian border.

From there, on 29 June 1941, began the attack on the left wing of the *Gebirgskorps* across the border toward Murmansk. In spite of all their efforts and sacrifices, the mountain troops were unable to decisively defeat the Soviet forces defending the frontier. Not only did they have to overcome a fiercely-determined enemy, they also had to contend with the forests and swamps of the tundra on the Polar Sea.

After taking Luosteri the *Gebirgsjäger* slogged their way to Titovka, where there was heavy fighting. Then the Liza was reached. The regiments and battalions had to dig in there to await supplies of weapons, food and materiel. The division had suffered considerable casualties which could not be made good. When September arrived, bringing longer, colder nights, the division crossed the Liza. But then the advance halted for good. Murmansk would never be reached.

The 6. Geb.D. arrived and in October 1941 relieved the 3. Geb.D. The only element of the division to remain in its positions was the Narvik

Regiment (Geb.Jäg.R. 139). The I./Geb.AR. 112 also stayed behind, however it became an army unit.

The 3. Geb.D. left the Polar Sea front for good and was transported or marched into the Vaasa area. From there it was transferred to the Grafenwöhr training grounds. The exhausted division was reformed in January 1942 and the following newly-established units were assigned to the 3. Geb.D.: Geb.Jäg.R. 138 and 144, Geb.AR. 112, Geb.Radf.Btl. 95, Geb.Pz.Jäg.Abt. 95 and Geb.Pi.Btl. 83.

The division remained under the command of the commander of *Wehrkreis XVIII* (Salzburg) until April 1942. Then it was transferred by rail and ship back to Norway, where the *Gebirgsjäger* were employed as training and security units in the areas around Lillehammer, Hamar. Elverum, Stören and Lake Mjösen.

Once declared a fully combat-ready division again, the 3. Geb.D. received orders to proceed to southern Russia to bolster the forces advancing into the Caucasus. The transports did not follow their planned course, however. Instead they were diverted into the eastern Baltic. The *Gebirgsjäger* were hastily disembarked in Reval, after which they proceeded on foot and in vehicles in the direction of Leningrad.

On 24 August 1942 the Russians had initiated the First Battle of Ladoga, attacking with 20 rifle divisions, four independent rifle brigades and five tank brigades. As soon as they arrived at the front before Gaitolovo the *Gebirgsjäger* were thrown into the front lines in an attempt to halt the enemy. With support from the 121. ID, the battle groups succeeded in closing a pocket around the Soviet VI Guards Corps, which had broken through the German front. This ended the First Battle of Ladoga.

The 3. Geb.D. was pulled out of the front and ordered back to Mga to prepare for transport to the southern front. The trains departed, but they never reached southern Russia. The trains halted west of Velikiye Luki (on the left wing of Army Group Center), and the *Gebirgsjäger* were forced to hastily disembark and set out on foot for the front lines south of Velikiye Luki. There they saw action in the hilly, marshy terrain around Novo Sokolniki, Chernosyan and Shirpina. First, however, several strong detachments had to be sent to the Millerovo area, where there was fierce fighting until March 1943.

The detached elements of the division returned in March 1943 and by the end of the month the 3. Geb.D. was a complete fighting unit again.

There was little time for training, for the division was finally sent to the southern sector of the Eastern Front.

In April 1943 the 3. Geb.D. detrained in the sector of the IV. AK. and subsequently took part in the positional fighting in the Voroshilovsk-Voroshilovgrad area. The main line of resistance was held until August, when orders were received to withdraw. There was confused fighting in the area of the XVII. AK. west of Kuibyshev, while battle groups from the division fought with the 304. ID. in the Redkino Gorge. At the end of August the division was transferred to the Kalmius, where there was fighting near Porkovo-Kireyevka.

Further confused fighting took place in September and October 1943 in the area Laspa-Ignatyevka, Volnovacha and southeast of Zaporozhye. From mid-September to the end of October the division withdrew via the "Wotan" and "Franken" positions into the Nikopol bridgehead. The division remained in positions there until January 1944. The division command post was located in Dneprovka southeast of Nikopol.

The new year brought further costly fighting withdrawals at the Ingulets and Dnestr under the command of the XXIX. and XXX. AK. The division suffered heavily, especially in the battle for Rascaeti, but was able to hold together. In July the division was transferred to the eastern Carpathians and was attached to an Hungarian corps headquarters of the German *8. Armee*. By September further fighting withdrawals had brought the *Gebirgsjäger* through the Bistritz Valley and over the Aurel Pass to the Maros and the Theiss. Withdrawal beyond the Gran River followed by the end of the year.

The last year of the war saw the men of the 3. Geb.D. in position south of the High Tatra, the regiments having received a brief rest and refit in the Rosenberg area. There followed positional warfare as part of the XXXXIX. Geb.K. between Jablunkau and Freistadt. The final weeks of the war found the *Gebirgsjäger* constantly on the retreat, through Olsa, Neutitschin and Mährisch-Weißkirchen, finally reaching the area south of Olmütz. On 8 May 1945 all fighting in this area ceased.

The Commanders:

Generalleutnant Dietl; *Generalmajor* Ringel; *Generalmajor* Kreysing; *Oberst* Mönch; *Generalmajor* Picker; *Generalleutnant* Wittmann; *Oberst* Pape; *Oberst* Kreppel; *Generalleutnant* Klatt.

4. Gebirgsdivision

Reacting to the first casualties suffered by the 2. and 3. Geb.D. in Norway, on 14 April 1940 the OKH ordered the formation of two further mountain divisions. In part, the units were supposed to be created from the elements left behind in Germany of the two formations deployed in the north. The rapid end of the western campaign meant that the two divisions were not formed, however; instead the remaining units of the mountain troops were assigned to a third mountain division, the 6. Geb.D.

The OKH order to form new armored and motorized divisions freed up a number of units, and on 23 October 1940 the 4. Geb.D. was formed at the Heuberg training grounds. The division was initially equipped as an airborne division, but was organized and armed like one of the other mountain divisions.

The new division headquarters was created from the Div.Stab. z.b.V. 412 of *Wehrkreis XII* (Wiesbaden). The regiments and division units came from the former 14., 19., 25., 27. and 36. ID. After the necessary renaming of units, the 4. Geb.D. consisted of the Geb.Jäg.R. 13 and 91; all other units bore the number 94.

After the necessary basic training, the new 4. Geb.D. transferred via Hungary, Romania and Bulgaria to the Yugoslavian border. There it was placed under the command of *Panzergruppe von Kleist.*

When the Yugoslavian campaign began on 6 April 1941, the task of the panzer group was to break through into southern Yugoslavia and take the city of Belgrade. The *Gebirgsjäger* attacked from the Grodec area and advanced toward Pirot. After smashing the initial enemy resistance they went over to the pursuit through Kujazeyac toward Krnsevac. After the cease-fire the division remained in the Kragujevac area southeast of Belgrade as part of the occupation force. Several weeks later, however, it was returned to southeastern Poland.

From there, on 22 June 1941, the *Gebirgsjäger* attacked the German Soviet frontier from the area around Rzeszow. After forcing a crossing of the San, they advanced through Grodek toward Lvov. The troops then advanced in the dust and heat into the Brzezany area. The Seret was crossed and the "Stalin Line" was breached near Derashnje. The subsequent advance in the direction of Vinnitsa resulted in fierce, costly fighting, at the end of which the city was taken. The 4. and 1. Geb.D. then closed the ring around Soviet troops at Uman.

The successful conclusion of the battle of encirclement in the Podvysokoye area was followed by an advance to the Dniepr. The broad river was crossed and the mountain division advanced toward the Sea of Azov. There was heavy fighting, especially near Malaya Belorsyorka, however the attacking forces reached the coast of the Sea of Azov near Mogila and Tokmak. The *Gebirgsjäger* then fought their way into the Stalino industrial region against stiffening resistance and took the city. Autumn arrived, bringing rain and mud, and the advance continued to and across the Mius River. There the troops occupied winter positions.

When 1942 began the division was in the Donets industrial region, fighting off fierce enemy attacks which often assumed threatening proportions. The battle for the so-called "Mius Position" went on for months until the start of the big German summer offensive to the Volga and the Caucasus.

At the end of July the division crossed the broad Don near and in Rostov. From there it marched beneath a blazing sun through the steppes and small villages of the Near-Caucasus. The division crossed the Kuban, advanced to Armavir and spearheaded the *XXXXIX. Gebirgskorps*'s advance toward the High-Caucasus. The *Gebirgsjäger* advanced up the high pass roads. The scene of snow-covered mountain peaks and deep valleys made up for the all the hardships to date. On 21 August 1942 a mixed battle group of the 1. and 4. Geb.D. climbed Mount Elbrus, the highest peak in the mighty Caucasus range.

The *Gebirgsjäger* found the Caucasian mountain world incomparably beautiful, however the snow-covered peaks and rock faces made the conduct of conventional warfare impossible. The fighting was restricted to a few pass roads which led to the coast. Then, on 22 August, just 28 kilometers from the coastal city of Suchum, the XXXXIX. Geb.K. was forced to abandon all hope of reaching the coast. Inadequate forces and supply problems had halted the advance. The *Gebirgsjäger* dug into the stony ground. Positional warfare began.

When the German withdrawal from the Caucasus began on 1 January 1943, the 4. Geb.D. left its mountain positions. The unit's heavy weapons and materiel were transported out, first on mules and then in trucks. Finally the *Gebirgsjäger* bade farewell to the graves of their fallen comrades.

The 4. Geb.D. joined the retreat of the German *17. Armee* into the Kuban region. In the months to come, this area of swamps and forests would be the battlefield for the *Gebirgsjäger*. As before, the division was under the command of the XXXXIX. Geb.K. on the left wing of the Kuban bridgehead.

On 2 February 1943 a large Soviet naval force landed three marine brigades in the bays around Novorossisk, initiating a four-week battle for the port city. In spite of all efforts, the enemy bridgehead could not be eliminated. The 4. Geb.D. suffered grievous losses in these battles, including the division commander, *Generalleutnant* Kreß. The 4. Geb.D. was one of the units which held out in the Kuban bridgehead to the very last.

The division transferred to the Crimea and was initially employed in the security role, especially around Stary Krim and Sevastopol. At the end of October elements were transported into the area around Melitopol and the Khersonyes bridgehead. Defensive fighting continued there until the end of the year. At the end of the year the German forces withdrew farther to the west. The troops stood and fought in the area around Vinnitsa, which they had captured in 1941. The *Gebirgsjäger* saw action between Olkhovets and Chishovka until February 1944.

The next two months brought the retreat to the Bug. Then the division participated in the relief attacks in support of the German divisions surrounded in the Cherkassy Pocket. This was followed by a further retreat into the area around Uman, also familiar from 1941. Fierce defensive fighting and a withdrawal to the Dnestr followed until July.

The 4. Geb.D. was pulled out of this sector of the front and in August 1944 was transferred to the Carpathians, where it participated in offensive and defensive actions at the Tatar Pass. In September followed a further retreat into Hungarian territory. The units fought a number of battles there and at the end of the year were in the Pelsoc and Rosenau area.

The last year of the war brought difficult fighting withdrawals and relief attacks for the beleaguered German and Hungarian divisions in the High Tatra. By March the *Gebirgsjäger* were already in Upper Silesia, still under the command of the XXXXIX. Geb.K. In April and early May the division participated in the battle for Tropau and the retreat behind the March. Finally the end of the division came between Brünn and Olmütz.

The Commanders:
Generalleutnant Eglseer; *Oberst* Wintergast; *Generalleutnant* Kreß; *Generalleutnant* Braun; *Oberst* Jank; *Generalleutnant* Breith; *Generalmajor* Bader.

5. Gebirgsdivision

Like the 4 Geb.D., the 5. Geb.D. was formed after the Western Campaign following the reorganization of infantry divisions into motorized or panzer divisions. At the same time, the 1. Geb.D. provided fully combat-ready units. The 5. Geb.D. was created under the responsibility of *Wehrkreis XVIII* (Salzburg) on 25 October 1940. Like the 4. Geb.D., the division initially bore the designation LL (airborne) after its name, however this was later dropped.

On 15 November 1940 the new mountain division was organized as follows:

Headquarters, Geb.Jäg.R. 85 (formerly with the 10. ID.) and 100 (formerly with the 1. Geb.D.), Geb.AR. 95 (I. Abt. formerly with the 1. Geb.D., III. Abt. formerly with the 10. ID.), Geb.Pi.Btl. 95 (elements formerly with the 10. and 18. ID.), Geb.NA. 95 (elements formerly with the 3., 10., 18., 19., 33. ID.), and Field Hospital 95 (formerly with the 33. ID.). By the summer of 1941 the division also received the Geb.AA. 95, Geb.Pz.Jäg.Abt. 95 and rear-echelon services with the number 95.

By the time the division reached full strength, however, it had already seen its first, costly action.

After training, in March 1941 the 5. Geb.D. transferred to Bulgaria as a so-called "Instructional Unit". From there it marched to the Greek border. When the Balkan campaign began in April 1941, the division moved out from the area surrounding Petritsch and breached the Metaxas Line. In the first days of the advance there was fierce fighting for key fortified points at Rupesko, Popoliwitsa, Istibi, Kelkaja and Arpaluki.

Then, as part of the XVIII. AK. of the *12. Armee*, the division began the pursuit of the shattered Greek army through Saloniki, Thermopylae and Lamia to the area of Athens. There the troops were granted several days of rest. Then, with little warning, they were transported by sea and air to Crete to help the parachute troops in their desperate struggle for the island.

The Geb.Jäg.R. 100 and the division headquarters landed at Maleme on the afternoon of 21 May. *Generalmajor* Ringel assume command of all the army and air force units in western Crete. The next day Geb.Jäg.R. 85 arrived, and while the 5. Geb.D. was not complete, together with the Geb.Jäg.R. 141, which arrived at the same time, it was a potent combat force.

The attack on Chanea began on 26 May 1941. The Geb.Jäg.R. 100 was positioned in the center of the attack front between Galatos and Karatos, while the 141st and 85th Regiments fought on the right wing. That same day the mixed force of British, Canadian, Australian and New Zealand troops were driven from Chanea and the city was occupied by mountain and parachute troops. Geb.Jäg.R. 100 was subsequently committed in pursuit of the fleeing enemy as far as Sfakia on the coast. Other elements reached the coast near Suda.

The 5. Geb.D. remained on Crete as an occupation unit until November 1941. Then it returned to the Salzburg area where the division was under the command of the Commander of the Replacement Army until the end of 1941. At this time the bicycle and anti-tank battalion had to be sent to the 3. Geb.D., which remained in Finland until the end of the war.

Suddenly, in February 1942, the division was loaded aboard transport trains of the Reichsbahn. The *Gebirgsjäger* traveled through all of eastern Germany, Latvia, Lithuania and Estonia to the northern wing of the Eastern Front. After detraining the division began practicing for an operation against the military port of Kronstadt.

The attack never took place, however, for at the beginning of 1942 the Red Army seized the initiative in the north, attacking from the Schlisselburg area as far as Demyansk in the sector of Army Group North. Instead of seeing action at Kronstadt, in the months that followed the 5. Geb.D. was employed wherever the front was ablaze.

The 5. Geb.D. fought at the Oranienbaum bridgehead, Lake Ladoga and, when the Soviets launched the First Battle of Ladoga on 24 August 1942, east of Sinyavino. The high ground bearing the same name became the focus of Soviet offensives in the following months. The division was later in the main line of resistance in the Mga area, on the Volkhov, and at the end of the year the *Gebirgsjäger* occupied positions on the Neva due east of Leningrad.

The division remained there until summer 1943. The troops had to battle an enemy whose strength grew month by month, but they also had to struggle against nature. They were plagued by snow and ice in winter, mud in the spring and swarms of biting insects in summer. At that time the 5. Geb.D. was under the command of the LIV. AK. of the *18. Armee*. When the Third Battle of Ladoga finally ended in September 1943, the division was forced to leave its main line of resistance and the battlefields of Army Group North.

By the time the 5. Geb.D. rolled through Germany again it was November. This time the trains rolled to Abruzzia. The division was one of the units sent to Italy to bolster German forces following that country's departure from the Axis alliance. The *Gebirgsjäger* found themselves in mountain country which suited them better than the swamps of northern Russia.

In May 1944 the Allied command issued orders for the two German armies fighting in Italy to be driven back to north of Rome and the gaining of the line Pisa-Rimini. During the first weeks of its deployment to Italy the 5. Geb.D. was under the command of the XXIV. PK. at the Monte Mare. From there began the withdrawal to Monte Croce, and from there via Cifalco to Monte Cassino, where the *Gebirgsjäger* took part in the fierce battle for the mountain. In May 1944 the division was transferred to the Adriatic Front east of Florence. There the division was under the command of the LXXVI. Pz.K. Together with two infantry divisions and the *1. Fallschirmjägerdivision* it helped defend the sector on the Mediterranean coast.

The Allied landings in southern France threatened the Alpine front on the French-Italian border. Therefore, in September 1944, the 5. Geb.D. received urgent orders to transfer into that sector. The Geb.Jäg.R. 100 was the first unit to reach the new sector and it was able to repulse the first enemy advances into the border region. Then the division units arrived. The *Gebirgsjäger* occupied a sector from the 3 843-meter Monte Viso to the Small St. Bernhard. At the beginning of October the front settled down into positional warfare after snow and ice made it impossible to conduct warfare.

In this last sector in its history, the 5. Geb.D. was under the command of the LXXV. AK of the *AOK Ligurien*. The main line of resistance was approximately 250 kilometers long and extended from Big St.

Bernhard in the north to Monte Viso in the south. In the words of its commander, at the beginning of 1945 the division was "the best-equipped division in the German Army." At that point it had more than 17,000 troops, 6,000 horses and mules and 1,000 motor vehicles.

The small war in the high mountains, for which the soldiers had been trained, went on until 25 April 1945. From that day on the undefeated division had to withdraw from its mountain positions and pull back on the right wing of the German forces in Italy into the Aosta Valley (right) and through Turin (left). All the while it engaged French and Italian partisans. The division withdrew to Turin in three march groups. There partisan groups and an Italian parachute brigade had barricaded the roads. The division lost 32 killed, 97 wounded and 140 missing in those final days. A cease-fire was negotiated by the Cardinal-Archbishop of Turin which held until the end of the war.

The Commanders:
Generalleutnant Ringel; *Generalleutnant* Schrank; *Generalmajor* Steets.

6. Gebirgsdivision

On 20 May 1940, with the western campaign in "high gear", the OKH ordered the formation of the 6. Geb.D. by *Wehrkreis XVIII* (Salzburg). The new division was to assemble at the Heuberg training ground by 6 June and be ready for operations by 10 June. Later it was decreed that the division was to be declared a peacetime division and was to be organized and equipped like the first three mountain divisions. The necessary units were taken from the 2. and 3. Geb.D. and the 69., 181., 209., 214., 239. and 557. ID.

Once the formation process was concluded, the division was organized as follows:

Div H.Q., Geb.Jäg.R. 141 and 143, Geb.AR. 112, Geb.AR. 118, Geb.Pz.Jäg.Abt. 47, Geb.Pi.Btl. 91, Geb.NA. 91, supply services, medical battalion and rear services with the number 91.

Soon after the start of the attack by Army Group C, the new 6. Geb.D. was sent across the Rhine and through the Maginot Line. On 17 June the units reached the area south of Freiburg and the next day crossed the river. Under the command of the XXV. AK., the battle groups of the division advanced through Schlettstadt into the St. Dié area. There they received news of the cease-fire.

The division remained on occupation duties in the area around Pontarlier until December 1940, when it was transferred to the XXXX. AK.mot.'s sector in Poland. Then, with little notice, in February 1941 there was another transfer. The division was sent to Romania and then at the end of March 1941 to the Greek border.

On 6 April 1941 the XVIII. Geb.K., of which the 6. Geb.D. was now part, attacked the Greek positions of the Metaxas Line from the area around Kolarewo and Gadrene. After breaching the Greek fortifications the *Gebirgsjäger* stormed toward Iraklion and the area of Doiren-Lake Butkowo. After the mountains had been crossed, the advance into the interior, via Lamia and Thebes toward Athens. A battle group from the division landed on Euböa.

After the campaign the division was halted and sent back to Germany. There, from May to August 1941, the division was brought up to full strength. With the division's fighting power restored, beginning on 22 August 1941 it was transferred to Finland. Part of the division proceeded Stettin-Oslo-Trondheim to Kirkenes, while the other part went Oslo-Vaasa to Rovanniemi.

In mid-October 1941 the complete division arrived in the Luostari area. From there the regiments and battalions moved into the Liza bridgehead and relieved the battle groups of the 2. and 3. Geb.D. The 3. Geb.D. departed the Polar Sea front and was transferred to the Eastern Front.

From then on the 6. Geb.D. was under the command of the XIX. Geb.K. and participated in that's corps' operations in the extreme north of Europe. The mountain troops' battleground was the wastes of Lappland on the extreme left flank of the entire Eastern Front.

For the division it was the beginning of a year-long battle which alternated between brief periods of offensive activity and months of defensive struggle. The division had to detach battle groups into the rear areas behind its front for weeks on end to counter Soviet commando groups which landed from the sea and attacked camps or supply troops. The only troops farther north were small battle groups of the army – including mountain units of *Oberst* van der Hoop on the Fischer Peninsula – and navy on the Kirkenes coastal sector.

The front remained unchanged until the autumn of 1944, when it became apparent that Finland was about to drop out of the war. As a precaution, the commanding AOK 20 ordered the withdrawal of all non-vital materials and rear-echelon services. The only route available for a possible evacuation was *Reichstrasse 50*, which had been built by Organization Todt (OT) personnel and Reich Labor Service (RAD) battalions.

"Operation Northern Light", the withdrawal of the XIX. Geb.K. from the Murmansk Front, began on 4 October 1944. Russian forces immediately gave pursuit and after just a few days had torn apart the front held by the 2. and 6. Geb.D. Consequently the division had to evacuate the Liza position, which projected eastward like a finger, sooner than planned. By doing so it avoided encirclement by the Russians.

Heavy blowing snow and fierce Russian artillery fire slowed all further withdrawal movements to the west. The Chief-of-Staff of the 20. Geb.Armee described the retreat:

> "After silencing the batteries barring the Petsamo Fjord, the Russians landed a battle group near Liinahamari under cover of fog and darkness. Russian tanks drove into the rear of the bridgehead still standing east of the Petsamojoki around Parkkina. The spearheads of an attack group operating from Luostari against the Tarnet Road appeared on the road. Command and troops mastered all of these ticklish situations. ..."

After the destruction of Kirkenes the troops reached the Finland-Norway border and thus escaped encirclement by the Soviets.

The 6. Geb.D. was employed in the security role in this area until the end of the year. In January 1945 it began withdrawing to the Lyngs Fjord. There the division occupied positions in the last line of security – the so-called "Sturmbock Position".

On the day of the surrender the last battle groups of the division surrendered to the British.

The Commanders:
Generalleutnant Schörner; *Generalleutnant* Philipp; *Generalmajor* Pemsel; *Oberst* Remold.

7. Gebirgsdivision

The division was formed by order of *Wehrkreis XIII* (Nuremberg) on 13 November 1941 from the former 99. lei.D. (99th Light Division) with the addition of *Gebirgsjäger* units from Bavaria, Styria, Carinthia and Tyrol.

The predecessor division (99. lei.D.) had been formed exactly one year earlier in the same *Wehrkreis* from elements of the existing 3., 46., 254. and 264. ID. The division took part in the 1941 eastern campaign as part of Army Group South and advanced through Zhitomir to Kiev. After its return to its home *Wehrkreis* it was transferred to the Grafenwöhr troop training grounds for reorganization as the new 7. Geb.D.

The former Regiments 206 and 218 became *Gebirgsjäger* regiments with the same numbers, while Artillery Regiment 82 became the new Geb.AR. 82. the remaining division units all received the number 99.

After the necessary changes in equipment and corresponding training, in January 1942 the division was split temporarily. One battle group – with most of the units – under the command of *Oberst* Krakau was transported to Finland and was deployed in the Hyrinsalmi area as "Battle Group Krakau". Another battle group under *Oberst* Hoffmeister, consisting of the Geb.Jäg.R. 206, III./Geb.Jäg.R. 218 and I./Geb.AR. 82, went to Army Group North where it was subordinated to various divisions as "Battle Group Hoffmeister". The rest of the division, including the division headquarters, remained at Grafenwöhr temporarily.

In the months that followed, "Battle Group Hoffmeister" was attached to various infantry divisions. Among others, the *Gebirgsjäger* fought with the 5. and 8. Jäg.D. and the 122. ID and participated in offensive and defensive actions at Lake Ilmen, near Koslovo, Kudrova, Jassvy and at the Redzy Road and the Gridino Road. Not until the summer of 1942 did the battle group leave the area of Army Group North. It was transported by ship from Reval to Finland.

Meanwhile "Battle Group Krakau" had relieved elements of the badly battered Finnish III Army Corps between Lake Pjä and Lake Top and became familiar with forest warfare in the primeval forests west of the White Sea. During the course of the varied combat operations the battle group was occasionally attached to the Finnish 11th Infantry Regiment.

In August 1942 the rest of the division arrived from Grafenwöhr, making the 7. Geb.D. a potent large unit once again. The German XVIII.

Geb.K. assumed command in this so-called "Kiestinki Sector". Under its command were the 7. Geb.D. and the newly-arrived *SS-Division "Nord"*.

From this point on the *Gebirgsjäger* fought mainly in small battle groups in the area east of Kienstinki against an enemy whose strength grew month by month. The forest battles in the area approximately 100 kilometers west of the Murmansk rail line were fought at places such as Lake Yeletyeoso and Lake Nizhnyeosoro northeast of Kiestinki, Lake Nyatovara and Lake Kangaswara, and the "Bunker Ridge" in the Senosero area. Sometimes the division fought side by side with the Geb.Brigade 139, which arrived on the Kiestinki front after covering 900 kilometers on skis from the coast of the Polar Sea by way of Rovanniemi.

The front due south of the Polar Sea in the Kiestinki and Uhtua area remained virtually unchanged until 1944. In September 1943 the FEB 99 (replacement training battalion) was reorganized as a ski battalion and was assigned the new number 82. The battalion was employed by the mountain corps as a mobile reserve and saw action on all fronts.

In the summer of 1944 the war-weariness of Finland was becoming increasingly apparent. With the consent of the OKW, the headquarters of the *20. Gebirgsarmee* worked out plans for the evacuation of German forces to take effect as soon as Finland declared its neutrality.

On 3 September 1944 the OKW launched "Operation Birch", which included a withdrawal south by the *Gebirgsarmee*. At that time the 7. Geb.D. was on the Armee's right wing in the Uhtua front. On orders, the XVIII. Geb.K. now had to pull out of the Kiestinki front. The 7. Geb.D. had to hold its position and not withdraw until the corps had reached the so-called "bulwark position" at the Sohjana narrows.

During the course of its withdrawal to the former Russian-Finnish border the 7. Geb.D. had only to fight a few rear-guard actions. The division's battle groups headed for Pudasjärvi, blowing bridges and mining roads. The Finnish villages were left untouched. The Finns remained neutral; not until the division had reached the Pudasjärvi did fighting begin with following Finnish units.

Two Finnish divisions, whose equipment included the latest German assault guns, tried to bar the 7. Geb.D.'s way to Rovanniemi. The Finns threw in two more divisions, ambushed German supply units and threatened the 7. Geb.D. with encirclement before it could reach Rovanniemi. Between Pudasjärvi and Rovanniemi the *Gebirgsjäger* battled the hard-pressing Finnish units. Nevertheless, the division reached Romanniemi and from there moved north along the border road (with Sweden). The division finally moved into the so-called "Sturmbock Position" on the Finnish-Norwegian border.

The battle groups of the division remained in this position until January 1945. The enemy no longer pressed forward. Beginning in February the position was abandoned and the division headed for Narvik. From there, beginning in March, the division withdrew south through northern and central Norway. Several battle groups reached Oslo. Then the war ended. The soldiers of the 7. Geb.D. were taken prisoner by the British.

The Commanders:
Generalleutnant Konrad; *Generalmajor* Weiß; *Generalleutnant* Martinek; *Generalleutnant* Krakau.

8. Gebirgsdivision

The parent unit of this division was the 157. Res.D. which had been formed in the autumn of 1942. Most of the units were formed from the area of *Wehrkreis VII* (Munich); the majority were reserve and replacement units from the Alpine areas of Bavaria. Soon after completing its training, the new division came under the command of the Commander-in-Chief Southwest. After completing the necessary training exercises, the division was committed against Maquis units appearing in the French Alps.

After the departure of Italy the effort shifted to combating Italian army units under Marshall Badoglio and later Italian partisan units set up by the communists. The division's main task was securing the mountain passes in the area of the French-Italian border, especially on Mont Cenis.

In September the division was briefly transferred into the Cannes area. Then it returned to the border region. There the division's mission was to guard the frontier from Mont Cenis to the Swiss border

Until the end of the year 1944 units of the division saw constant action in the area around Grenoble, at Mont Blanc, Mont Cenis and the tunnel of the same name near Modane. In those months the former reserve division was reorganized into a reserve mountain division.

The two *Gebirgsjäger* regiments received the numbers 296 and 297, the mountain artillery regiment the number 1057. The same number was allocated to the formerly independent mountain pioneer battalion and all remaining division elements with the prefix "*Gebirgs-*". In October 1944 the OKH ordered the division renamed the *8. Gebirgsdivision*; thus the former reserve division had become an active division.

The division was now withdrawn from its positions, where it had been subordinated to the LXXV. AK. of the *Armee Ligurien*. At the beginning of 1945 the new 8. Geb.D. transferred into the area around Bologna. There the three infantry divisions of the XXIV. Pz.K. had occupied the main line of resistance in front of the city. The 8. Geb.D. was inserted between the 94. and 65. ID. The corps was part of the *14. Armee*, which was supposed to repulse the British 8th Army's assault between Lake Commachio on the Adriatic and the southwards-facing salient in front of Bologna.

The 8. Geb.D. fought bitterly in defense of Marzobotto, Vado and Scasoli and on both sides of the Setta Valley. The troops held their positions for months, until on 14 April 1945 the American 5th Army launched an offensive against Bologna. Hundreds of close-support aircraft and hundreds of guns opened the offensive. There was no defense against them. Five divisions of the American II Corps attacked the 8. Geb.D.'s main line of resistance. The *Gebirgsjäger* fought bitterly for one day but then had to abandon their positions.

In the ten days that followed, the German *14. Armee* collapsed. Bologna was abandoned without a fight and the army withdrew to the north. The division's battle groups defended themselves desperately in the battles for Sasso Marconi. The heavy weapons were lost and the unit was encircled. Elements were able to break out to *Strada Nationale 64* and assembled around San Giovanni. The division carried out a series of fighting withdrawals in which one unit after another was captured by the enemy.

Remnants of the 8. Geb.D. withdrew to the Panaro and the Po. Now there no longer was a homogenous unit. Troops crossed the river near Ostiglia by climbing over the blown railway bridge or by floating across on wooden beams. Many soldiers and all the weapons remained behind. The end of the 8. Geb.D. had come.

Commander:
Generalleutnant Schricker.

9. Gebirgsdivision

With the agreement of the Commander-in-Chief of the *Luftwaffe*, at the end of March 1945 the Commander of the Replacement Army ordered the formation of a new *Gebirgsdivision* to be designated the 9. Geb.D. Responsible for the new formation were *Wehrkreise VII* (Munich) and *XVIII* (Salzburg). The following units were combined to form the new *Gebirgsdivision*:

Army Non-Commissioned Officer School for Mountain Troops in Wörgl;
Mountain Artillery School Dachstein;
SS-Gebirgsjäger-Ersatz-Btl. Loeben (SS Mountain Replacement Battalion);
elements of *Gebirgsjäger* Replacement Battalions 137 Salzburg and 139 Klagenfurt;
personnel of the disbanded *Kampfgeschwader 27 "Boelcke"*;
units of the *Volkssturm* from Styria.

Since the unit was a combination of officers, non-commissioned officers and enlisted men from the army and air force, the selected division emblem was a mountain pick-axe crossed with an aircraft propeller.

After a few days of training the battle groups of the division were moved into the area around the Semmering where they soon engaged Soviet armored forces advancing on Vienna. Lacking heavy weapons, the troops were soon forced to give ground and withdrew into the area of Liezen an der Enns.

There, on the day of the surrender, the war ended for the 9. Geb.D.

Commander:
Oberst Raithel.

10. Gebirgsdivision

The last *Gebirgsdivision* was formed in May 1945, within days of the surrender, and received the number 10, however the new division had existed since autumn 1944. At that time it was designated Div. z.b.V. 140 (Special Purpose Division 140). Under its command were independent brigades and army troops which could be committed by the Geb.AOK 20 as the situation required.

When the Div. z.b.V. 140 was formed it included the Geb.Brigade 139, which was created from the former Geb.Jäg.R. 139 which had occupied Narvik in April-May 1940. The regiment had remained in northern Norway where it was enlarged to brigade size and formed the core of the division.

The new division was under the command of the XVIII. Geb.K. and during the corps' retreat to the Norwegian border in November 1944 it was expanded into Division Group Kräutler with support from the 7. Geb.D.

The withdrawal took the division group into the Oulu area, where there was heavy fighting. After a brief rest there followed defensive fighting at the Kemijoki in the Kemi area. The unit subsequently withdrew through the Kemijoki Valley to Rovanniemi. Further defensive actions followed in the border region and in December 1944 the division group was in the Narvik area.

At the start of 1945 the Div. z.b.V. 140 consisted of headquarters, Geb.Jäg.Brig. 138, Geb.Jäg.Btl. 3 and 6, Headquarters AR. 931 with II./ Geb.AR. 82 and I./Geb.AR. 112 and division units.

The division marched from Narvik through central and southern Norway into the Mosjoen area. There, on 5 May 1945, it was renamed the 10. Geb.D., three days before the war ended.

Commander:
Generalmajor Kräutler.

188. Gebirgsdivision

The only *Gebirgsdivision* with a "high house number" was formed in October 1943 under the command of *Wehrkreis XVIII* (Salzburg) as the *188. Geb.Ausbildungs-D.* (Mountain Training Division). From their personnel, Geb.Jäg.Ers.Rgter. 137 and 139, both located in the *Wehrkreis*, provided the Geb.Feld-Ausbildungs-Rgter. 901 and 902 (Mountain Field Training Regiments). The Geb.Ers.AR 112 (Mountain Artillery Replacement Regiment) formed the Geb.Feldausbildungs-AR. 112 with a single battalion. The remaining division units were provided by the other units of the Ers.D. 418 (Salzburg).

The role of this, the only mountain training division, was to train the regiments, battalions and independent battalions under its command in occupied territory until they were combat ready. Soon after its formation the division was transferred to Istria in the area of the Commander-in-Chief Southwest. The 188. Geb.Feld-Ausbildungs-D. was under the command of the LXXXXVI. and LXXXXVII. AK.z.b.V. and thus the commander of the Adriatic coast. After training and unit exercises, the units were employed to guard supply lines of communication and military facilities. This mission was expanded when partisans became more active in Istria after the start of the big Soviet offensive on the Eastern Front.

The division was now required to provide battle groups to track down and engage these partisan units. These missions often resulted in painful casualties, since the troops in training only remained with the division for a short time before being sent to an active unit and were thus inexperienced.

The division remained in Istria. In September 1944 it transferred to the area in and around Trieste and assumed responsibility for coastal defense there.

The division remained in that area, but as the Allies approached northern Italy it was subordinated directly to the AOK 10 of Army Group C. In February 1945 the division, which was still referred to and led as a reserve division, was sent into the front. In May 1945 the 188. Geb.Feldausbildungs-D. was forced to lay down its arms in the Trieste area.

Commander:
Generalleutnant von Hößlin.

Appendices:

Gebirgstruppe Command List, 10 December 1944

Unit	Commander	Chief-of-Staff or Ia
20. Gebirgs-Armee:	Generaloberst Rendulic	Generalmajor Hölter
XV. Geb.K.:	Gen.d.Pz.Tr. Fehn	Oberstleutnant Einbeck
XVII. Geb.K.:	Gen.d.Inf. Hochbaum	Oberst Jais
XIX. Geb.K.:	Gen.d.Geb.Tr. Jodl	Oberst Purucker
XXI. Geb.K.:	Gen.d.Inf. von Leyser	Oberst von Klocke
XXII. Geb.K.:	Gen.d.Geb.Tr. Lanz	Oberstleutnant Geitner
XXXVI. Geb.K.:	Gen.d.Geb.Tr. Vogel	Oberstleutnant Schmidt
XXXXIX. Geb.K.:	Gen.d.Geb.Tr. von Le Suire	Oberst Haidlen
LI. Geb.K.	Gen.d.Geb.Tr. Feuerstein	Oberst Gartmayr
1. Geb.D.:	Generalleutnant Kübler, J.	Oberst Failmüller
2. Geb.D.:	Generalleutnant Degen	Oberstleutnant Roschmann
3. Geb.D.:	Generalleutnant Klatt	Oberstleutnant Hartmann
4. Geb.D.:	Generalleutnant Breith	Major Brandner
5. Geb.D.:	Generalleutnant Schrank	Major Pröhl
6. Geb.D.:	Generalleutnant Pemsel	Oberstleutnant Vogel
7. Geb.D.:	Generalleutnant Krakau	Oberstleutnant Paumgarten
157. Geb.D.:	Generalmajor Schricker*	Major Koepper
188. Geb.D.:	Generalleutnant von Hößlin	Hauptmann Mittelwallner

(* later 8. Geb.D.)

Wearers of High-Ranking Decorations

Knight's Cross with Oak Leaves and Swords

No.	Rank, Name, Position	Date of Award
53	Gen.d.Geb.Tr. Kreysing, C.O. XVII. AK.	18/4/1944
72	GenOb. Dietl, C.I.C. 20. Geb.Armee	1/7/1944
122	GenOb. Rendulic, C.I.C. 20. Geb.Armee	18/1/1945

Knight's Cross with Oak Leaves

1	GenLt. Dietl, Com.Gen. Geb.K. "Norwegen"	19/7/1940

164	Hptm. Von Hirschfeld, C.O. II./Geb.Jäg.R. 98	23/12/1942
183	GenLt. Kreysing, C.O. 3. Geb.D.	20/1/1943
222	Obfw. Schlee, Plat.Com. 6./Geb.Jäg.R. 13	6/4/1943
312	GenLt. Ringel, C.O. 5. Geb.D.	25/10/1943
316	Obstlt. Count von der Goltz, C.O. Geb.Jäg.R. 144	2/11/1943
398	Gen.d.Geb.Tr. Schörner, ComGen. XXXX. PK. (former commander XIX. Geb.K.)	17/2/1944
686	GenLt. Klatt, C.O. 3. Geb.D.	26/12/1944

Naming of Divisions in OKW Communiqués 1943-45

1. Geb.D. on 10/10/1944
3. Geb.D. on 24/10/1943; 30/3; 4/10; 5/11/1944; 14/2/1944
4. Geb.D. on 12/3; 28/9/1944
5. Geb.D. on 12/8/1943
6. Geb.D. on 15/11/1944
7. Geb.D. on 24/8/1943

2

Fallschirmjäger (Parachute Troops)

Pre-war History

On 29 January 1926 the Reich Minister of Aviation and Commander-in-Chief of the *Luftwaffe* issued an order for a number of personnel from the "*General* Göring" Regiment to be trained as instructors in parachute jumping and packing. All were to be volunteers. This order was in effect the first step in the formation of the German parachute corps, which, nine years later at the end of the Second World War, fielded eleven divisions.

The first requirement for the formation of such a novel fighting force was the creation of a parachute school where future parachute troops could be trained in jumping and all that went with it. The initial site chosen was Neubrandenburg air base, however the choice ultimately fell on Stendal air base which was located near the small village of Borstel. The first German parachute school was created there in 1936.

The officers, non-commissioned officers and enlisted men who volunteered for this mission were breaking new ground. With the exception of the USSR, no other state possessed a parachute corps. There were no reference points from which to proceed and no foreign experience to evaluate.

The only exception was the Soviet Union. The Red Army had begun forming a parachute corps in 1928. The sport of parachuting became very popular in Russia in the 1920s. With government support it had evolved into a recognized people's sport. The numerous sport jumpers were brought together in the "Osviachim Association" which maintained branches across the country. When the German-Russian war broke out in 1941 there were more than 1,000 such branches in the Soviet Union.

Marshall Tukhachevski, then Commander-in-Chief of the Red Army, was an especially ardent supporter of the new Russian parachute corps. The first action by the Russian parachute troops took place during maneuvers in 1930. A young lieutenant and his platoon jumped south of Moscow and took the target, a high-ranking headquarters, completely by surprise. This was the first use of parachute troops in military history. The experts were perplexed, however, when in 1936 the Soviets lifted the veil of secrecy that had surrounded their parachute corps. This happened during large-scale maneuvers in the Ukraine, to which foreign military attaches and other dignitaries had been invited as observers. 1,000 Soviet parachute troops landed out of a clear blue sky, followed by 5,000 Red Army soldiers in transport machines. The world had never seen anything like it.

One of the observers, the future British General Wavell, wrote to his government: "I would not have believed such an operation possible if I had not witnessed it myself." Five years later General Wavell, then British Commander-in-Chief in the Near East, would see first hand the effectiveness of the new weapon when German parachute troops attacked the island of Crete from the air.

The commanders of the German armed forces (*Wehrmacht*) were surely aware of the existence of Russian parachute troops when, in 1936, the order was issued for soldiers to receive parachute training. But for reasons that are well-known, the Germans had to make their own way and develop their parachute corps from the ground up, technically and tactically.

The *Oberkommando der Wehrmacht* (Armed Forces High Command) and the *Oberkommando der Luftwaffe* (Air Force High Command) had naturally done preliminary work. This ultimately led to the order of 29 January 1936 from the Reich Minister of Aviation and Commander-in-Chief of the *Luftwaffe* (L.A.Nr.262/36 g III, Ia) which resulted in the formation of a German parachute corps. Formations concerned with airborne landing missions had existed in both the German air force and the army for a year, however 29 January 1936 must be seen as the birth date of the German paratrooper (*Fallschirmjäger*).

On 1 April 1935 the (Prussian) Police Force Unit *"General Göring"* had received its new military designation *"Regiment General Göring"*. At the end of September in the same year, the regiment's commander, *Oberstleutnant* Jakoby, received the following order from *General* Göring in Rominten, East Prussia: "Effective 1 October 1935 the *General Göring* Regiment will be transferred to the *Luftwaffe*. Volunteers from the regiment will form a parachute battalion which will be the foundation of the future German parachute corps..."

During the month of October the regiment was transferred to the Altengrabow training grounds, there to develop and test the organization and operational possibilities of a parachute regiment. At the end of the regiment's stay at Altengrabow it witnessed a parachute jump at the Döberitz training grounds. That same day 600 officers, non-commissioned officers and enlisted men volunteered for the parachute corps.

These volunteers formed the *I. (Fallschirmschützen-) Bataillon* (1st Parachutist-Rifleman Battalion) of the *Regiment "General Göring"*. The battalion adjutant was *Oberleutnant* Vogel, signals officer *Leutnant* Dunz. The four companies were commanded by *Oberleutnant* Walther, *Oberleutnant* Kroh, *Oberleutnant* Schulz and *Hauptmann* Reinberger. When the Regiment "*General Göring*" was reorganized in November 1935 the *I. (Fallschirmschützen-) Bataillon* was renamed *IV. Bataillon.* (On 1 April 1938 it left the Regiment "*General Göring*" and became *I./ Fallschirmjäger-Regiment 1.*) Almost all of the instructors of the Stendal Parachute School came from this battalion.

The old city of Stendal with its brick Gothic, located 60 kilometers north of Magdeburg in Altmark, had become the cradle of the German parachute corps. The first German parachute school was created in 1936 at the Borstel air base located a few kilometers north of the city. Stendal later became the peacetime base of the *Fallschirm-Infanterie-Bataillon* (parachute-infantry battalion).

Over the years tens of thousands of paratroopers were trained at the Stendal parachute school, later also in Wittstock, Brunswick and Chateaudun (France). Serious accidents were very rare, and only a few parachutists lost their lives in training jumps. Such painful experiences occurred only during the young corps' developmental years, however.

Nevertheless, the parachutist was conscious of the fact that there was a certain element of danger in every jump, and he had to overcome his fears before each jump. Consequently, to last in this corps a soldier had to possess courage and determination. *General* Student wrote:

> "In dealing with the paratrooper one must take into consideration his pride in being a member of the parachute corps. One must be more broad-minded, caring and comradely than elsewhere. Development must be based far more on trust than on discipline and obedience. The paratrooper must act independently without orders."

Only those men who volunteered for service in the parachute corps and who had been cleared by the doctor as "fit for parachute duty" were allowed to jump. The fact that all German parachute troops were volunteers was one of the decisive factors behind the combat effectiveness and stamina they displayed during the Second World War.

One of the first technical questions that had to be answered was the use of a suitable parachute. True, the parachute had been in use by aircraft crews as a life-saving device for many years, however that type of parachute was not suitable as a "means of transport" for parachute troops. The rescue chute was operated manually and a relatively long drop time was required from activation of the release handle until full deployment. Consequently the parachute's minimum safety altitude was 200 meters. A manual parachute was thus out of the question for low-altitude operations. What was required was a parachute which opened automatically and which required a drop of at most 50 meters to deploy fully. The RZ 1 automatic parachute was developed in a short time. It was later replaced by the improved RZ 16 and RZ 36.

The prospective paratrooper first encountered the parachute early in his training, not in jumping but in chute packing practice. For each paratrooper had to learn how to pack a parachute before making his first jump with a parachute he had packed himself. Preparation for the first jump required intensive training in all aspects of the jump: preparations in the aircraft, the jump, behavior in the air and – most importantly – on landing. The landing was the most difficult part of the jump. It was vital that the parachutist not injure himself on landing. A paratrooper who broke an arm or a leg or suffered a concussion was lost for the post-landing action.

In strong winds it was difficult for a parachutist to gain his feet after landing, as the wind-blown parachute dragged him downwind. But the paratrooper was trained to cope with this situation. If he was unable to get to his feet he would use his folding knife to cut the parachute lines and release himself.

The average training time at the parachute school was eight weeks. During this time candidates completed six jumps. Soldiers who were not equal to the physical or mental demands or those who displayed serious character flaws or were hesitant to jump, were dropped from the course and transferred to another branch of the service.

Progress in establishing the Parachutist-Rifleman Battalion was such that by the summer of 1937 it already had three rifle companies, a machine-gun company and a pioneer company.

In addition to creating a paratrooper unit, the *Luftwaffe* High Command also had to form an air unit to fly for the paratroopers. The first step taken was to convert *Behelfskampfgeschwader 172* (Auxiliary Bomber Wing 172), which was a reserve unit for the possibility of mobilization, into an air unit.

On 1 October 1937 the *IV. Gruppe* of the new *Kampfgeschwader 152 "Hindenburg"* under *Major* Ziervogel was renamed *Kampfgruppe z.b.V. 1* (Special-Purpose Bomber Group 1). This designation implied that the *Gruppe* was earmarked exclusively for the new parachute corps. The *Gruppe*, which was equipped with the Ju 52/3m, transferred to Fürstenwalde air base, from where it was to practice its new and specialized role.

Meanwhile, on 5 November 1936 the Commander-in-Chief of the *Luftwaffe* had authorized the *Fallschirmschützenabzeichen* (Parachutist's Badge). The badge was worn by all members of the new corps who had completed at least six practice jumps. On receiving the badge the soldier also received the parachutist's certificate. Each parachutist also received a pay bonus of 65 to 120 *Reichsmark* depending on his rank.

For egotistic reasons the Army High Command had also begun plans for its own parachute corps. The army turned these plans into reality soon after the occupation of the Rhineland.

Formation of a parachute-infantry company was begun at Stendal. The company was supposed to be equipped with heavy machine-guns and heavy mortars. Formation of the company was largely complete by the autumn of 1937, allowing it to participate in the large-scale *Wehrmacht* maneuvers in Mecklenburg. Command of the *Fallschirm-Infanterie-Kompanie* (Parachute-Infantry Company) was in the hands of Oblt. Zahn, assisted by Oblt. Pelz.

The first step in the formation of an army battalion was the transfer of *Major* Heidrich to the *Fallschirm-Infanterie-Kompanie*. Heidrich had formerly been a tactics instructor at the Officer Candidate School in Potsdam. At the age of 41 he completed jump training at Stendal just like the young volunteers. Then, in the spring of 1938, he participated in the formation of the *Fallschirm-Infanterie-Bataillon*, becoming its first commanding officer.

The *Luftwaffe* High Command was quicker and more determined in its efforts. The new corps was not planned as an independent arm and was part of the *Fliegertruppe*. Consequently its soldiers wore yellow collar patches. But in its ultimate form adopted in 1938 the *Fallschirmtruppe* received its own ordnance inspectorate in the *Reichsluftfahrtministerium* with the title *L in 11* and was brought into the newly-formed *7. Fliegerdivision*.

On 1 July 1938 *Generalmajor* Student assumed command of the German *Fallschirmtruppe*, at least those elements which were part of the *Luftwaffe*. Student soon placed his personal stamp on the young corps. To all intents and purposes he became the organizer of the *Fallschirmtruppe*. After encountering German parachute troops, the Allies formed their own airborne and parachute units, following the very same training principles which Student had devised.

Generalmajor Student received his initial military training at the Potsdam and Lichterfelde cadet schools. He served in the *Yorckschen Jäger* until 1913 and then transferred to the then new *Fliegertruppe* (air corps). In 1914 he went into the field and by 1916 he was already commander of *Jagdstaffel 9*, part of Army Group "German Crown Prince". After the First World War Student was taken into the 100,000-man army. There, from 1919 to 1928, he was the Official Responsible for Aviation Technology in the Defense Ministry (in the *Reichsheer*, which had no air arm!). He was thus responsible for all technical developments from materials to complete aircraft. In this position he undertook numerous fact-finding missions to the states of Europe, especially the Soviet Union.

On 1 September 1933 this highly-capable officer, now with the rank of *Oberstleutnant*, was transferred to the newly-created State Ministry of Aviation (*Reichsluftfahrtministerium*). He encountered parachute troops as commander of the Rechlin and Travemünde test stations. His links with the new corps became closer after he was made Inspector of *Luftwaffe* Training Schools, which placed the parachute school in Stendal under his command.

On 1 July 1938 *Generalmajor* Student was given command of the German parachute corps. This marked the beginning of the rational expansion of this new arm and the development of clear operational principles.

The new Inspector of Parachute Troops could not understand why there should be two completely separate units within the *Wehrmacht* performing the same role. The *Fallschirm-Infanterie-Bataillon* remained under the command of the army, however, even though it was temporarily placed under *Generalmajor* Student's tactical control during the occupation of the Sudetenland in 1938. Then, on 1 January 1939, it was transferred to the *Luftwaffe* as the *II. Bataillon* of *Fallschirmjäger-Regiment 1*. In the summer of 1938 the unit's commander *Major* Heidrich, who had been selected to command *Fallschirmjäger-Regiment 2*, the next parachute regiment to be formed, was transferred to the headquarters of the *7. Flieger- (Fallschirm)-) Division*. His successor was Hptm. Prager.

In summer 1938 the battalion had transferred from Stendal to Brunswick.

The next units to be formed were *III./Fallschirmjäger-Regiment* 1 (Hptm. Schulz) and *I./* (Hptm. Noster) and *II./Fallschirmjäger-Regiment* 2 (Hptm. Pietzonka) at Gardelegen and Tangermünde near Stendal. *Generalmajor* Student's most pressing task was to create a common training program for the parachute units with their various backgrounds.

The parachute battalion commanded by *Major* Bräuer (*Luftwaffe*), which had been designated "Parachute-Destroyer Battalion", was organized, equipped and trained to destroy important targets behind enemy lines. In general terms its operational principles mirrored those of the Allied commandos in the Second World War. The battalion was organized into individual "destroyer squads" which were to be committed by parachute against bridges, supply and communications centers, headquarters and production facilities.

The army's small parachute force, on the other hand, was earmarked for purely tactical missions within the framework of localized operations. Its principle roles included seizing and holding important sectors of terrain and potential bottlenecks, the surprise capture of bridges and establishing of bridgeheads, and attacking fortifications from the rear. *Generalmajor* Student's concepts went much farther. He rejected both the "destroyer tactic" and the army's tactically limited view of the parachute corps' role. Just as forward-thinking army commanders had turned the panzer arm into an independent offensive weapon, very early on *Generalmajor* Student recognized the great strategic potential of the parachute corps. Like the panzer arm, it was capable of achieving wide-ranging strategic goals. It was the *Wehrmacht*'s fast corps with a degree of mobility previously unimagined. In Student's view, the parachute corps' ability to operate independently while exploiting the element of surprise was its greatest strength.

As the training of the parachute units progressed, further air transport units were formed. By 1 September 1938 there were five independent bomber groups equipped with the Ju 52, which was suitable for the transport of men and materiel. The designation "air transport unit" did not exist as yet. The bomber units were all still designated "*z.b.V.*" (special purpose). They were:

Kampfgruppe z.b.V. 1 at Fürstenwalde
C.O.: *Oberst* Morzik
Kampfgruppe z.b.V. 2 at Briest (Westhavelland)
C.O.: *Oberstleutnant* von Lindenau
Kampfgruppe z.b.V. 4 at Tutow
C.O.: *Oberstleutnant* Leesemann
Kampfgruppe z.b.V. 5 at Fassberg
C.O.: *Oberstleutnant* Alefeld
Kampfgruppe z.b.V. 6 at Lechfeld
C.O.: *Major* Krahl

The German air command's concepts for the transport of parachute troops represented a leap forward compared to previous military thinking. It was believed that pinpoint targets could be taken by surprise through the use of unpowered aircraft. As a result, development work by the German Research Institute for Gliding Flight in Darmstadt-Griesheim was expanded greatly. In 1937 designer Jacobs produced the DFS 230 transport glider which seemed suited to such tasks.

The training of selected pilots and later also members of the *Fallschirmtruppe* began the same year and was pursued more intensively at Fürstenwalde in 1938. In September an operational detachment was formed there and was transferred to Prenzlau.

The "Transport Glider Detachment" under the command of Oblt. Kieß was formed in 1938 under the cover designation *17./Kampfgeschwader z.b.V. 5* and was equipped with 65 DFS 230 gliders. Each glider was capable of accommodating ten men.

Meanwhile training of the parachute troops was proceeding at a rapid pace, and the parachute school at Stendal (commander *Oberstleutnant* Bassenge) was frequently visited by high-ranking military personnel eager to see for themselves the state of training. As commanders of branches of the *Wehrmacht*, *Feldmarschall* Göring and *Generaloberst* Baron von Fritsch were frequent guests of the training school.

The first official appearance by the young corps took place in 1937. During the harvest festival on the Bückeberg men of the Army Parachute Company under the command of Oblt. Pelz jumped from their Ju 52s in public for the first time. During subsequent autumn maneuvers, which were attended by Italian head of state Mussolini, army paratroopers carried out jumps "behind enemy lines".

On 1 January 1939 *Fallschirmjägerregiment* 1 – whose headquarters was not established until 1 June 1939 – formed a *III. Bataillon* from volunteers. Under the command of *Oberstleutnant* Bräuer, all twelve companies took part in the big parade to mark Hitler's birthday on 20 April 1939. The paratroopers paraded in their new uniforms wearing the special helmets created for them and their jump gear. This parade had a great impact, especially abroad.

The political conflicts which developed between Germany and the western powers in 1938 as a result of the German occupation of the Sudetenland generated a surprise discussion concerning the new corps' first military mission.

Fallschirmjägerregiment 1 was involved in mobilization preparations for the military buildup by the *Wehrmacht* against Czechoslovakia. Among its assigned roles was a drop behind the fortified border areas in the Ore Mountains and Moravia to forcibly open the advance roads for the army.

Like the paratroopers, the air transport units and the glider detachment were also placed on alert.

After a peaceful settlement was reached, the units returned to their garrisons and settled back into their routines…

…until on 26 August 1939 the *Wehrmacht* High Command ordered the secret mobilization.

The Second World War

The OKW ordered the "secret mobilization" on 26 August 1939. Following receipt of the order, the units of the *7. Fliegerdivision* transferred to airfields around Breslau, from where they were to be deployed as the military situation developed. Two operational orders arrived. The first was for a drop into the Graudenz area, however East Prussian soldiers captured the fortress before the order to take off was given. The second order was for a drop near Pulawy to seize the bridge over the Vistula. By the time the Ju 52s took off, German tanks were already on the bridge.

Not until mid-September was a mission order received for the entire division. The companies were placed aboard trucks and subsequently occupied several airfields and bridges between the Bug and the Vistula in order to bar the retreat of the Polish army trapped in the corridor. The parachute troops fought two battles: III. Btl. engaged Polish troops near Sucha and on 24 September II. Btl. fought a pitched battle near Wola Gulowska. In both instances the German side suffered casualties.

In mid-October 1939 the division returned to its peacetime bases. In the weeks that followed, preparations for the western campaign, in which the headquarters of the *7. Fliegerdivision* became involved early on, were marked by intensive jump and combat training. Specially selected officers, non-commissioned officers and enlisted men (11 officers and 427 men) were moved to the Hildesheim area to form *"Sturmgruppe Koch"*, named for its commander, Hptm. Koch. Together with the glider detachment commanded by Oblt. Kieß, the parachute troops underwent special training.

Contrary to expectations, however, western Europe was not the next theater of war.

"Operation Weserübung", the occupation of Denmark and Norway, was not to *Generalmajor* Student's taste. He believed that deploying his units piecemeal would reduce the corps' striking power.

Early on the morning of 9 April 1940 the Fallsch.Jäg.R. 1 took off in Ju 52 aircraft. One company successfully stormed the big bridge between the Danish islands of Falster and Fünen. Another company seized the major airport at Stavanger-Sola. Twenty-nine Ju 52s carrying two more companies had to break off their approach to Fornebu airfield near Oslo on account of fog. Because of unexpectedly stubborn resistance, poor roads and even poorer weather, the campaign in Norway dragged on longer than expected. The parachute troops were called on once again after British forces landed near Namsos and, three days later, near Andalsnes. The 1./Fallsch.Jäg.R. 1 was dropped near Namsos in poor weather. Only a few officers and *Jäger* succeeded in linking up. They fought the British for three days then were forced to surrender.

The northern theater (Narvik) produced another jump mission in mid-May 1940. The I./Fallsch.Jäg.R. 1 took off in several waves and, operating from Trondheim, saw action at Narvik on several days. On 8 June parachute troops were the first German troops to enter the city and occupy Narvik for the second time.

By that time, however, the western campaign had entered its second phase, a campaign which would confirm the significance of this newest German corps.

According to the plans of the OKH, the *7. Fliegerdivision*, operating in close coordination with army troops, was to capture the line of forts on the Belgian border in order to open the way into the interior for the panzer divisions.

The *Fallschirmjäger* had to take out the vital fort of Eben Emael on the Albert Canal in the initial hours of the western campaign so as to enable the *6. Armee* to break through the Belgian system of fortifications.

Early on the morning of 10 May 1940 a force of 53 Ju 52s, three He 111s and 50 DFS 230 gliders took off from the Cologne-Ostheim and Cologne-Butzweilerhof airfields carrying "*Sturmgruppe Koch*" (consisting of 11 officers, 427 non-commissioned officers and enlisted men and 50 glider pilots). The group was divided into assault teams "Steel" (Oblt. Altmann), "Concrete" (Lt. Schacht), "Iron" (Lt. Schächter) and "Granite" (Oblt. Witzig).

Each of the four teams had its own target. The "Granite" team (2 officers, 28 non-commissioned officers, 56 enlisted men and 11 pilots) was sent against the strongest fort, Eben Emael. Assault team "Steel" was committed against bridges over the Albert Canal near Veldwezelt, "Concrete" the bridges near Vroenhoven, and "Iron" the bridge in Canne.

The drops were made on schedule, however the surprised Belgians regrouped quickly. They succeeded in blowing the bridge in Canne, however the others were taken by the parachute troops and held against furious counterattacks by the Belgians. At 1600 hours East Prussian infantry and pioneers linked up with the "Iron" team and a short time later established contact with "Concrete". Contact was not established with the "Steel" group until the next day.

The battle for Eben Emael raged on. While the *Fallschirmjäger* were able to occupy the plateau, the Belgians continued to defend the fort. Not until 0700 hours the next day were army assault engineers able to link up with the parachute troops. Eben Emael surrendered at 1315 hours on 11 May.

The way into Belgium had been opened for the *6. Armee*.

The second area of operations for the parachute troops was Holland. With the agreement of the OKH, the OKL formed an airborne landing corps consisting of the *7. Fliegerdivision* and the *22. Infanterie-Division*. The attack order for the corps had already been issued on 22 February 1940. It began with the following words:

> "1. Intention: Seizure of Fortress Holland from the rear, outflanking the Grebbe Line.
> 2. Execution: Airborne Landing Corps (Fl.D. 7 and 22. ID.) to be dropped in the Moerdijk-Dordrecht-Rotterdam-Hague area with the objective of:
> a) holding open the bridges near Moerdijk (main effort) and Rotterdam for following fast troops of the army, which will be sent in via Tilburg-Breda (Fl.D. 7), and
> b) occupying the capital of The Hague and immediately eliminating the crown, government and armed forces command located there (22. ID.) …"

Kampfgeschwader z.b.V. 1 flew into Dutch airspace early on the morning of 10 May 1940. On board were all three battalions of *Fallsch.Jäg.R. 1* and *I./Fallsch.Jäg.R. 2*. The *I./Fallsch.Jäg.R. 1* and the regimental headquarters were dropped near Tweede Tool between

Dordrecht and Moerdijk. *II. Btl.* landed on both sides of the Moerdijk bridges. All were seized except the railway bridge in Dordrecht. *III. Btl.* landed on Waalhaven airfield near Rotterdam under heavy enemy fire. *I./ Fallsch.Jäg.R. 2* was also supposed to land there. The Dutch defenders fought with amazing courage.

Only then did Ju 52s fly in the first units of the *22. ID.*, which was supposed to take Rotterdam. The landing came at a high cost in men and materiel. Rotterdam withstood the initial onslaught. *Generalmajor* Student, who had jumped with his parachute troops, set up his command post in Rijksoord. He was dismayed to discover that his units only had contact with the *III./IR 16* in Rotterdam. Transport aircraft had landed the other two battalions of *IR. 16* in Waalhaven, however, and relief was on its way to the parachute troops.

The parachute troops holding the bridges in Moerdijk and Dordrecht were relieved by tanks of the *9. Panzer-Division* on 13 May. The daring operation by the German parachute troops was over, for the next day Rotterdam surrendered after a fearful bombing raid on the city by German aircraft. *Generalmajor* Student was wounded in the head on the final day of the campaign.

Twenty-seven officers and non-commissioned officers were decorated with the Knight's Cross during and soon after the western campaign. The parachute corps thus received ten percent of all the Knight's Crosses awarded in the western campaign.

Just one paratrooper, Oblt. Schmidt, commander of 1*./Fallsch.Jäg.R.* 1, was awarded the Knight's Cross in the Norwegian campaign. He was awarded the decoration for his soldiers' efforts at Dombas.

When the fighting was over the *Fallschirmjäger* returned to their peacetime garrisons. This was followed by the corps' second expansion phase. The *Fallsch.Jäg.R. 2* was expanded into three battalions and the experimental battalions stationed in the Friedrichshafen area were combined to form the *Luftlande-Sturmregiment 1* (Airborne Assault Regiment 1) with a strength of four battalions. The fame gained by the parachute corps in Holland brought in a flood of volunteers and the *Fallsch. Jäg.R. 3* was formed in the late summer.

The new *7. Fliegerdivision*, now under the command of *Generalleutnant* Student, had learned from its initial combat experiences. The troops received new weapons, including a 105-mm recoilless rifle which was dropped from transport aircraft beneath five parachutes.

Earlier weapons were modernized based on the corps' experiences in Holland. It received the 37-mm anti-tank gun and the 80-mm mortar, both of which could be air-delivered. The weapons and equipment were packed in canisters and dropped by parachute.

In the winter of 1940-42 the OKL formed the Corps Headquarters of the *XI. Fliegerkorps* under the command of *General der Flieger* Student. Under the corps headquarters, in addition to the general staff, were the *Luftlande-Sturm-Regiment* (*Generalmajor* Meindl), *Ln-Abt. 41 (mot.)* (Hptm. Hoppe) plus the *7. Fliegerdivision* (*Generalleutnant* Süßmann) with the *Fallsch.Jäg.R. 1* (Oberst Bräuer), *Fallsch.Jäg.R. 2* (*Oberst* Sturm), *Fallsch.Jäg.R. 3* (*Oberst* Heidrich), *Pi.Btl.*, *Art.Abt.*, MG-Btl. and *Pz.Jäg.Abt.*, each with the number 7.

The *Wehrmacht* was in the midst of preparations for the invasion of the Soviet Union in spring 1941 when the Italian government "threw a wrench into the works". The Italian attack against Greece, launched from Albania, soon bogged down and Rome asked Berlin for help.

The result was the German attack on Yugoslavia and Greece which began on 6 April 1941. During planning for the campaign, at the beginning of March the *Fallsch.Jäg.R. 2* was transferred to the Plovdiv area of Bulgaria to prevent British landings on Greek islands.

The *Wehrmacht* crushed the Yugoslavian and Greek armies in a few weeks and forced the British Expeditionary Corps back in the direction of the coast. The British withdrew toward the Peloponnes. *Fallsch.Jäg.R. 2* received combat orders. At 0500 hours on 26 April 1941 transport aircraft took off from Larissa carrying two parachute battalions, engineer, signals and medical companies and a battery of artillery of the *Fallsch.Jäg.R.* 2. The parachute troops were dropped on both sides of the Corinth Canal, where they established bridgeheads. Even though British artillery fire destroyed the bridge, the parachute troops did not give up. They fought their way to the Corinth airfield and in the face of bitter resistance stormed toward Nauplion. They were there when the army's advance guard arrived on the morning of 28 April.

On that same day orders were issued for the parachute troops to go into action elsewhere: Crete. On 20 April *General der Flieger* Student had proposed to the OKL that the island be taken. One day later he described his plan to Hitler, who was enthusiastic. The German high command had no idea of the strength of the enemy forces on the island, however, and his led to the heavy losses suffered by the German paratroopers, especially after the British command became aware of the German plan on 15 May from decoded radio messages.

The German plan called for simultaneous landings by the *7. Fliegerdivision* from the air and the 5. Geb.D. from the sea. Since 1 October 1940 the *7. Fliegerdivision* had been led by *Generalleutnant* Süßmann, a former police officer who had transferred to the *Luftwaffe* in 1935, subsequently commanding several *Kampfgeschwader* before becoming an officer in the Reich Ministry of Aviation (RLM).

For the assault on Crete the *7. Fliegerdivision* was divided into three groups:

Group West: objective Maleme airfield
Group Center: capital of Chanea, Suda Bay and Rethymnon airfield
Group East: city and airfield of Iraklion.

The limited number of available transport machines (439 Ju 52s) made it necessary to launch the attack in two waves. The first wave was to occupy the Maleme airfield, Chanea and Suda Bay early on 20 May. The aircraft would return to base and then transport the second wave of paratroopers to their objectives, the airfields at Rethymnon and Iraklion. Bombing raids by units of the *VIII. Fliegerkorps* would begin the assault on Crete. These would be immediately followed by the *7. Fliegerdivision*'s airborne landings.

This largest, most successful and costliest operation was to be the high point in the history of the German parachute corps.

While the Ju 52s dropped two battalions of the *Luftlande-Sturm-Regiment* around the Maleme airfield, gliders carrying I. Btl. landed at the Tavronitis bridge near Maleme, on Suda Bay and on Hill 107. At the same time *Fallsch.Jäg.R.* 3 jumped onto the plain southwest of Chanea. Losses were heavy everywhere, as the paratroopers were dropped right onto the enemy positions. *Generalleutnant* Sußmann was himself killed when his glider crashed on the island of Ägina.

Only seven of the Ju 52s which took off were lost, however after returning to the Greek islands there was a lengthy delay caused by refueling and a sandstorm. As a result, it took much too long to get the 1st and 2nd Regiments to Crete.

Until the evening of 20 may the German paratroopers on Crete fought for their very survival. It was not until that evening that *General* Student in Athens began to develop an approximate picture of how the situation was developing. He ordered the capture of Maleme and its airport for the next day.

Oberst Ramcke was one of the first to jump that day and he assumed command over all the parachute troops in the Maleme area. Late in the afternoon on 21 May the Maleme airfield finally fell into German hands. At the same time the *Geb.Jäg.R. 100* of the 5. Geb.D. began landing. By evening it had three battalions ashore. *Generalmajor* Ringel assumed command of the German forces around Maleme. At this time *General* Student, commander of all German forces on Crete, also arrived.

The attack on the capital city of Chanea now began. Parachute and mountain troops stormed the enemy positions and on 27 May they entered the capital city of Crete. In practical terms, the battle for Crete was now decided.

The OKW's final report concluded with the sentence:

> "With this the fighting came to an end, and a glorious new page was added to the history of the German armed forces."

On 19 August 1941 the Supreme Commander received the parachute corps' Knight's Cross recipients at *Führer* Headquarters. That day he said to *General* Student:

> "Crete has shown that the days of the parachute corps are over. The parachute arm is purely a weapon of surprise. The surprise factor has now been exhausted."

The youngest branch of the German armed services had been dealt a temporary death blow; from this point on it would be used in the infantry role.

By autumn 1941 the units of the *7. Fliegerdivision* had been brought up to strength again and were ready to return to action – but as trench fighters, not paratroopers. The campaign against the Soviet Union had not developed as the German command had hoped. The Red Army was stronger than had been anticipated and had inflicted heavy losses on the German army and air force.

It was at this point that the *Wehrmacht* command remembered the paratroopers. The *II./ Luftlande-Sturm-Regiment* was the first unit to be sent to the Eastern Front, being assigned to the force which was supposed to reduce the Soviet bridgehead at Petrushino near Leningrad. During the course of this first ground operation the paratroopers suffered enormous losses, including all their officers.

At the end of September 1941 the headquarters of the *7. Fliegerdivision*, now under the command of *Generalmajor* Petersen, a former army officer, and almost all of its units were sent to the northern front. In two months of fierce fighting the *7. Fliegerdivision* lost 3,000 men killed or wounded, more than in the Battle of Crete.

When the year 1942 began, the division was still in the northern sector of the Eastern Front, although several units were at the Mius River in the extreme south of the Eastern Front and one battalion of the assault regiment was in the central sector, while another was fighting at Rzhev. Thus it was that in the first months of the new year the sole German parachute division was scattered all over the huge Russian front.

Then in April 1942 *General der Flieger* Student received orders to report to the Commander-in-Chief South in Rome. The OKW and the Italian command were planning the conquest of Malta. The Italian *Folgore* Parachute Division, which had been trained by *Generalmajor* Ramcke, was to play a central role in the operation, but Hitler rejected every plan of this type.

Back in Germany, *General* Student began reorganizing the units of his division as they returned. Two new regiments were formed: *Fallsch.Jäg.R. 4* from volunteers and *Fallsch.Jäg.R.* 5 from the *Luftlande-Sturm-Regiment*. The units received more modern weapons and a new type of parachute. The Ju 52s were replaced as glider tugs by converted Ju 87 dive-bombers and the new large-capacity Me 321 glider had become available.

In August 1942 the paratroopers were called into action again. The German *Panzerarmee Afrika* had been forced to halt its offensive into Egypt and was now on the defensive. The Ramcke Brigade, named for its commander, was hastily put together and flown to Tobruk. The brigade consisted of four parachute battalions, one parachute artillery battalion, an anti-tank company and a pioneer company.

In August 1942 the Ramcke Brigade was positioned on the panzer army's extreme right flank in the Quattara Depression in the midst of the desolate, sun-scorched desert. There, in spite of the heat, dust and sand, the men held out against a superior enemy until the order to retreat was issued on 2 November 1942. The paratroopers fell back on foot, dragging their heavy weapons. After four days a British vehicle column crossed their path. The paratroopers stormed the column, capturing trucks loaded with fuel, water and food. The next day the now motorized Ramcke Brigade reached the German front near Marsa Matruk after a 350-kilometers trek through the desert.

Generalmajor Ramcke, who had volunteered for the parachute corps in 1940 as a 51-year-old infantry *Oberst* and who parachuted into Crete in 1941 to assume command of all the parachute troops there, was awarded the Knight's Cross with Oak Leaves for his brigade's performance in North Africa. He thus became the first member of the parachute corps to receive this high decoration.

By this time the parachute corps had been restored to full strength in Germany and awaited new orders. These came in October 1942. The mission was not an airborne operation in the Caucasus, as had been expected, but employment in the infantry role in the central sector of the Eastern Front. The regiments, battalions and independent battalions remained there until spring 1943.

The only unit not in action on the Eastern Front was the *Fallsch.Jäg.R. 5*, which since 14 November 1942 had been in Tunisia/North Africa. The Allied landings in Morocco five days earlier and their rapid advance into the rear of the German panzer army in Tunisia made it necessary for the Germans to occupy Tunis.

The *Fallsch.Jäg.R. 5* landed the *Fallsch.Jäg.Pi.Btl. 21* around Tunis in two days and, under the command of *Major* Koch, it defended a front of 300 kilometers. On 26 November the paratroopers, by then under the command of *Oberst* Barenthin, engaged British and American armored forces for the first time. The *Fallschirmjäger* held their front, defending every meter of ground, until the fighting in Tunisia ended on the Bon Peninsula on 12 May 1943.

During the winter of 1942-43 *General* Student was able to have all parachute units pulled out of the Eastern Front and transferred to France. There the parachute corps, with the exception of the units fighting in Tunisia, was to be reorganized.

The units of the *7. Fliegerdivision* which were transported to France by May 1943, plus additional units from Germany, were combined into a new division which received the number 1. Created at the same time as the Fallsch.Jäg.Div. 1 was a second division with the number 2. The Fallsch.Jäg.Div. 1 consisted of Regiments 1, 3 and 4 plus division units with the number 1; the Fallsch.Jäg.Div. 2 consisted of Regiments 2, 6 and 7 plus division units with the number 2.

The *1. Fallsch.Jäg.Div.* was stationed in the Avignon area, the *2. Fallsch.Jäg.Div.* in the area of Nimes. In command of the 1st was *Generalmajor* Heidrich, of the 2nd *Generalmajor* Ramcke. The two divisions did not have to wait long for operational orders, for on 9 July 1943 a large Allied force landed on the island of Sicily.

The result was the creation of a new theater of war in which the German parachute troops would play a major role until the end of the war.

The *1. Fallsch.Jäg.Div.* received operational orders on the day after the landings. The first unit to depart for the front was the *Fallsch.Jäg.R. 3*, which was flown to Rome, from where it proceeded to Sicily. The regiment jumped south of Catania on the afternoon of 12 July. The *Fallsch.Jäg.R. 4* followed the next day along with other battalions of the division.

The *Fallschirmjäger* fought a tough battle against British paratroopers which had also been dropped onto the Plain of Catania. The first German regiment to land was surrounded but managed to fight its way out after two days.

For the next two years the Italian Theater remained the main area of operations of the parachute corps which, in spite of heavy casualties, fought with great élan on every battlefield. The following lists depicts some of the areas of operations in the year 1943:

1. Fallsch.Jäg.Div.:	Sicily until 17 August; Apulia until 2 September; Calabria until the end of September, including Battle of Salerno; Abruzzia until November.
2. Fallsch.Jäg.Div.:	Rome and surroundings from 26 July; occupation of the Aegean islands until

	November: freeing of the Duce in September. Transferred to the Eastern Front in October.
4. Fallsch.Jäg.Div.	Formation in the Perugia area from left-behind elements of the *2. Fallsch.Jäg.Div.* and volunteers from two Italian parachute divisions.

Beginning in October 1943 a name began dominating the military events in Italy, one which would also become synonymous with the German parachute corps: Monte Cassino.

The Allied advance on Rome had been halted in front of the German "Gustav Line". Located in the north of the Liri Valley, the 516-meter Monte Cassino was a strategic point on the road to Rome. The Germans planned to spare the monastery at Monte Cassino, the cultural-historical center of the entire region. On 14 October 1943, therefore, the *Fallschirmjäger-Instandsetzungs-Bataillon 1* (Parachute Repair Battalion 1) began removing the monastery's cultural treasures. During the next weeks, 120 trucks transported 700 crates containing valuable books, manuscripts, documents and other cultural treasures to the Vatican.

The evacuation came just in time, for on 6 December the Allies began pounding Monte Cassino with a barrage that was reminiscent of Verdun in the First World War. British, American and French batteries began ranging in on 3 January 1944. The Allied attack began two days later.

The German parachute troops were committed on 6 February 1944, when the German front appeared ready to crack. The *Fallschirmjäger* occupied Monte Cassino and stormed Calvary Mountain, which had been occupied by the Americans. The battle lasted five days and ended in favor of the German defenders.

The Second Battle of Monte Cassino began on 15 February 1944, when Allied bombers dropped 453 tons of bombs on the mountain. The Benedictine monastery was destroyed. Five days later *Feldmarschall* Kesselring ordered the *1. Fallsch.Jäg.Div.* to defend Monte Cassino. *Generalleutnant* Heidrich's soldiers held their positions. The second battle ended on 28 February, the third began on 15 March. When this battle ended at the end of March 1944 the German parachute troops had earned the name "Green Devils". The Allied Commander-in-Chief, General Alexander, wrote to the British Prime-Minister:

> "I doubt that there is another body of men in the world which could endure what these people have and subsequently fight on with the same determination."

At 2300 hours on 11 May 1943 the Allies opened up with the heaviest artillery barrage seen in one night in the Italian Theater. When morning came, Allied bombers flew 2,991 sorties against the German positions; it was the start of the Fourth Battle of Monte Cassino. By the third day of the battle the *1. Fallsch.Jäg.Div.* was down to just 700 fighting men, but it was the only large unit which held its positions. Then, on 17 May, the order came to abandon the position. That night the paratroopers left the city of Cassino, the mountain of the same name and the monastery.

In the meantime, however, the Allies had begun a major sea borne landing in the bay of Anzio-Nettuno. The first battalions of the newly-formed *4. Fallsch.Jäg.Div.* had been in action there since January. Until the end of February the division defended the extreme right sector of the German front, right on the Tyrrhenian Sea. By the beginning of June 1944 the superior Allied forces were able to break out of the bridgehead and begin the offensive against Rome and central Italy.

Both parachute divisions were now in action, and after abandoning the Italian capital they took part in the defense of the "Gothic Line". This was followed by the defensive battle on the Adriatic Coast and the defense of Florence, after which they withdrew into the mountains. Both divisions remained in northern Italy until the war ended there at the end of April – beginning of May.

The second theater in which the German parachute troops demonstrated their fighting ability was in the west following the Allied invasion of France.

The year 1944 brought a reorganization of the German parachute corps. In January the OKL formed the I. Fallsch.K. (I Parachute Corps) from the II. Luftw.Feld.K. (II Luftwaffe Field Corps) which had previously been deployed on the Eastern Front. The new corps was sent to Italy. At the same time, the II. Fallsch.K. was formed from the XIII. Fl.-K. (XIII Air Corps) in France and in May was transferred to Normandy. One month later the High Command formed the 1. Fallsch.Armee (1st Parachute Army) commanded by *Generaloberst* Student. The most urgent task facing the new army headquarters was the organization of further expansion of the parachute corps.

Three new division-size units (5., 6. and 7. Fallsch.Jäg.Div.) were created. All were established and trained in the west and saw action there following the Allied invasion on 6 June 1944.

The 2., 3., 5., 6., and 7. Fallsch.Jäg.Div. served on the Western Front until the end of 1944. On the day of the invasion the Fallsch.Jäg.Rgt. 6 became the first parachute unit to engage the enemy, near Carentan. The parachute troops defended the small Norman town for five days. Shortly thereafter, the 3. Fallsch.Jäg.Div. went into action near St. Lô and fought there until mid-July when the city and the combat area had to be evacuated after Allied armor broke through the German front at Avranches.

It was this point that the 2. Fallsch.Jäg.Div. came to prominence. In 1943 the division saw action in the southern sector of the Eastern Front. In April it was withdrawn from the Kishinev area to Cologne to rest and refit. From there it was rushed to Brest on the Atlantic Coast. The 2. Fallsch.Jäg.Div. defended the fortress against superior American and French forces until 19 September 1944. *Generalleutnant* Ramcke, division commander, was awarded the Knight's Cross with Swords and Diamonds, becoming the most highly-decorated member of the German parachute corps.

The remaining parachute divisions deployed in the west took part in the difficult fighting withdrawal to the borders of the Reich. The High Command of the 1. Fallsch.Armee, which commanded all units on the Dutch-Belgian border, including the 1. Fallsch.Jäg.Div., was able to defeat the British airborne operation "Market Garden" at Arnhem. After the battle the newly-formed II. Fallsch.K. (7. and 8. Fallsch.Jäg.Div.) was placed under the *Armee*'s command.

In mid-December 1944 the German armed forces tried to regain the initiative on the Western Front. "Operation Watch on the Rhine" went into the history books as the Ardennes Offensive. Three German armies were allocated the task of smashing the Allied front in the Ardennes and driving via Dinant to Antwerp. The attack force included the 3. and 5. Fallsch.Jäg.Div. and Fallsch.Jäg.Rgt. 6.

The 3. Fallsch.Jäg.Div. was part of the 6. SS-Pz.Armee, which was positioned on the right at the point of main effort. The 5. Fallsch.Jäg.Div. was part of the 7. Armee which guarded the left wing of the attack front.

The independent Fallsch.Jäg.Rgt. 6 took part in the last jump mission by German parachute troops in the Second World War. It was supposed to be dropped over the Hohen Venn and open the way for the I. SS-PzK.

Oberstleutnant von der Heydte's regiment was carried by Ju 52 transports, however only about half reached the target area. Of the 870 paratroopers, only about 450 jumped and approximately 100 assembled on the ground. It was a hopeless undertaking and the survivors surrendered near Monschau.

The other two parachute divisions committed in the ground battle halted their advance on 22 December. The 3. Fallsch.Jäg.Div. had become bogged down in front of the Elsenborn troop training grounds in the first days of the attack. The 5. Fallsch.Jäg.Div. was the only division which could keep pace with the armored forces and advanced farther to the west than any other unit of the *7. Armee*.

1945, the last year of the war, saw the formation of four more parachute divisions, although they lacked the spirit and fighting ability of the older divisions. The OKH's situation maps from the final months of the war place the divisions in the following areas of operation:

1. Fallsch.Jäg.Div. (*Generalmajor* Schulz): combat in the Bologna area and retreat into the mountains.

2. Fallsch.Jäg.Div. (*Generalleutnant* Lackner): reformed in the Amersfoort area on the Lower Rhine, retreat through Arnhem and Wesel, elements in the Ruhr pocket.

3. Fallsch.Jäg.Div. (*Generalleutnant* Schimpf): fighting withdrawal from the Ardennes to the Eifel, across the Rhine to Remagen, and in the Ruhr pocket.

4. Fallsch.Jäg.Div. (*Generalleutnant* Trettner): combat in the Bologna area and retreat into the mountains.

5. Fallsch.Jäg.Div. (*Generalmajor* Heilmann): retreat from the Ardennes through the Eifel to the Ruhr pocket.

6. Fallsch.Jäg.Div. (*Generalleutnant* Plocher): combat in

	the Arnhem area, retreat to the Lower Rhine, final actions between Kleve and Wesel.
7. Fallsch.Jäg.Div.	(*Generalleutnant* Erdmann): rushed transfer from the Pfalz to the Lower Rhine between Kleve and Geldern, final actions in Ostfriesland.
8. Fallsch.Jäg.Div.	(*Generalmajor* Wadehn): formation in Wesel area in February, final actions between Ems and Weser.
9. Fallsch.Jäg.Div.	(*Oberst* Herrmann): February formed in the Stargard area of Pomerania, action on the Eastern Front, retreat to Berlin.
10. Fallsch.Jäg.Div.	(*Oberst* von Hoffmann) formed in Krems/Donau area in April, elements to Moravia and the Alpine region.
11. Fallsch.Jäg.Div.	(elements only, no commander): formation in the Oldenburg area in March, transferred to the Netherlands in April, no action.

The German parachute corps ceased to exist when the German armed forces surrendered unconditionally on 8 May 1945.

Senior Commands

Armeeoberkommando. 1. Fallschirm-Armee

In March 1944 the OKL formed the *AOK 1. Fallsch.Armee* from the corps headquarters of the *XI. Fliegerkorps* and *Armeegruppe "Narwa"* of the army. All parachute corps and parachute divisions came under the command of the new parachute army high command. After the invasion in the west it was employed as a front-line army high command which also had army units under its command. The AOK 1. Fallsch.Armee assumed command of all front-line units in the Netherlands and Lower Rhine region.

Commander-in-Chief (OB):
Generaloberst Student; *General der Fallschirmtruppe* Schlemm (from November 1944); *General der Infanterie* Blumentritt (from March 1945); *Generaloberst* Student (from beginning of April 1945); *General der Infanterie* Straube (from end of April 1945).

Corps Headquarters of *Fallschirmjägerkorps* (Fallsch.K.):

No.	Formed	Commanding General
I.	1944	*General der Fallschirmtruppe* Schlemm,
		General der Fallschirmtruppe Heidrich.
II.	1944	*General der Fallschirmtruppe* Meindl.

The formation of corps headquarters did not take place until January 1944 after the increase in the number of divisions. The objective was to combine these into corps-size combat units and employ them in combat, something which the war situation prevented from happening.

Formed in October 1940, Headquarters, *XI. Fliegerkorps*, which was commanded by *General der Flieger* Student and which had command of the individual divisions, was not a front-line command. From the corps headquarters emerged the *Fallschirmtruppe*'s AOK 1.

Gen.Kdo. I. Fallsch.K. (Headquarters, I Parachute Corps) was created from the headquarters of the *II. Luftw.Feld.K.* (II *Luftwaffe* Field Corps) which had previously commanded the *Luftwaffe* field divisions deployed in the central sector of the Eastern Front (Vitebsk area). The *Gen.Kdo. II. Fallsch.K.* was formed from the *Gen.Kdo. XIII. Fl.K.*

Fallschirm-Panzerkorps "Hermann Göring"

At the end of 1942 the OKL began forming units which, while they were to belong to the *Luftwaffe*, were intended purely for the ground combat role. These included the so-called "*Luftwaffe* Field Divisions" and the later *Fallschirm-Panzerkorps "Hermann Göring"*. Like the *Reichsführer-SS*, the Commander-in-Chief of the *Luftwaffe* was to have ground combat units under his direct command. This marked the beginning of a three-way division of German land forces – those commanded by the OKH, plus units under the command of the OKL and the *Reichsführer-SS*. As the war went on, however, it was found that these units had to operate under army commands, corps and armies.

In 1942 the OKL ordered the formation of the *Fallschirm-Panzerdivision "Hermann Göring"* in the Berlin area. The necessary units were taken from the existing *Regiment "General Göring"* (the guard unit of the *Luftwaffe* Commander-in-Chief) and from all *Luftwaffe* training units and air regiments plus volunteers. Tanks, motorized vehicles, weapons and equipment were provided by the army. The first commander of the new division was the former commander of the *Wachregiment "General Göring"*.

After training and exercises the new panzer division was transferred to Army Group D in southern France. There the unit was quartered in the Bordeaux area and continued training. After the surrender of the German forces in Tunisia the unit was ordered to Italy. The elements of the division arrived in the Caserta area of southern Italy in May 1943 and were attached to the *10. Armee* as a reserve unit.

Four weeks later the *Panzer-Division "Hermann Göring"* was moved to Sicily, making it the only German division on that strategic island (the 15. Pz.Gren.Div. was also there but was still in the process of being formed).

Allied forces landed on Sicily during the night of 10 July 1943. The division engaged Allied troops that night and soon found itself in distress. Even after the *1. Fallsch.Jäg.Div.* parachuted into Sicily two days later, the German forces were able to offer little resistance to the superior Anglo-American forces.

Following the evacuation of Sicily, the *Fallschirm-Panzerdivision "Hermann Göring"* moved into Calabria and in September 1943 was on the right wing of the German front facing Salerno. Following the Allied breakout the division joined the general German retreat toward the Volturno.

The next major action took place in January 1944 when the Allies landed near Nettuno. Units of the division were in the thick of the fighting at the Anzio-Nettuno bridgehead until the end of February. Having been bled white in the fighting at Anzio, the division was transferred by Army Group Command C to Tuscany to rest and refit. It subsequently took part in the fighting for Rome. The division again suffered severe casualties and, acting on orders, withdrew to Lake Trasimeni, where the front was stabilized.

The division was then relieved and transported to the Eastern Front, where in August it saw action in the Baranov bridgehead. The division became familiar with defensive actions against the superior forces of the Red Army in the Warsaw area and during the subsequent retreat to East Prussia. There the *Fallschirm-Panzerkorps "Hermann Göring"* was placed under the command of the *Korpskommando Fallsch.Pz.K. "Hermann Göring"* together with the newly-formed *Fallsch.Pz.Gren.D. "Hermann Göring"*.

The result, in October 1944, was the single parachute-panzer corps formed from the *Fliegerkorps XI*. The corps assumed command of the two divisions which bore the name "Hermann Göring". In addition to the panzer division there was now also a panzer-grenadier division, which was also made up of *Luftwaffe* personnel plus some army officers, non-commissioned officers and technicians.

Both divisions were placed in the front line in the Gumbinnen area of East Prussia. There they fought fierce, costly defensive actions in October and November. The two divisions fought grimly against the superior Soviet forces which had encircled the *4. Armee* in East Prussia. Together with the Panzer-Grenadier-Division the corps headquarters fought its way back to the Heiligenbeil pocket, from which the HW and the remnants of the division escaped by sea to Pomerania.

The *Fallschirm-Panzerdivision "Hermann Göring"* was rushed to southern Poland at the end of 1944-beginning of 1945. It detrained in the Lodz area and following the start of the last big Soviet offensive fought its way back through Breslau to Saxony.

Meanwhile, the corps command and the remnants of the shattered panzer-grenadier division had also arrived there, and the units of the *Fallschirm-Panzerkorps "Hermann Göring"* fought side by side again.

After being brought back up to strength, the divisions fought with the *4. Panzerarmee* in the Lauban-Bautzen area. The last German success at Bautzen was followed by a fighting withdrawal across the Elbe into the area of Dresden. There, on the day of the surrender, the divisions blew up their last remaining vehicles and weapons. The troops were taken prisoner by the Soviets and few survived to return home.

The Commanders:

Fallschirm-Panzerkorps "Hermann Göring":
Generalleutnant Schmalz

Fallschirm-Panzerdivision "Hermann Göring":
Generalmajor Conrath;
Generalmajor Schmalz (from September 1943); *Generalmajor* von Necker (from October 1944); *Generalmajor* Lemcke (from February 1945).
Fallschirm-Panzer-Grenadier-Division "Hermann Göring":
Generalmajor Walther.

The Divisions

7. Fliegerdivision

At the start of the war the *Luftwaffe* was organized into four *Luftflotten* (air fleets) which commanded a total of nine *Luftgaue* (administrative commands), one *Luftgaukommando* (administrative command headquarters) and six *Fliegerdivisionen* (air divisions). The *7. Fliegerdivision* (the parachute corps) remained under the direct command of the OKL (*Luftwaffe* High Command). The *7. Fliegerdivision*'s actions in the period 1939 to 1943 is described in detail in the previous chapter "The Second World War".

The actions of the German parachute corps in the early years of the war were those of the *7. Fliegerdivision*.

It was not until the parachute corps had been assigned a costly ground role that *General der Flieger* Student was able to convince his superiors that the parachute corps was not an elements of the infantry. After their returned from the scattered fronts where they had been deployed, the regiments, battalions and independent battalions were once again combined into a parachute corps and thus the history of the *7. Fliegerdivision* ended in February 1943.

Commanders:
Generalleutnant Student; *Generalleutnant* Süßmann (from October 1940); *Generalleutnant* Petersen (from June 1941).

1. Fallschirmjägerdivision

The piecemeal employment of the various units of the *7. Fliegerdivision* in all sectors of the Eastern Front was absolutely contrary to the operational principles of the parachute corps. Furthermore, the large number of men of all ages volunteering for the parachute corps proved that there was still a great deal of interested in the force.

In the winter of 1942-43 *General* Student finally succeeded in having the battalions of his corps withdrawn from their positions between Leningrad and the Mius. By February 1943 these units had been transported to troop training grounds and army camps in southern France, Normandy and Brittany. In the months that followed, the first three new, potent parachute divisions were formed under the *I.* and *II. Fallsch.K.*, likewise newly-formed.

The *1. Fallsch.Jäg.Div.* was formed in southern France by May 1943 and was organized as follows:

Headquarters,
Fallsch.Jäg.R. 1, *3* and *4*, each with three battalions,
Fallsch.AR. 1 with two battalions,
Pz.Jäg.Abt. 1 with six companies,
Fallsch.Flak-Abt. with one battery,
Fallsch.Pi.Btl. 1 with four companies,
Fallsch.MG-Btl. 1 with three companies,
Fallsch.NA. 1 with two companies,
Fallsch.Vers.R. 1 with various companies.

The division's first commander was a veteran officer, *Generalmajor* Heidrich.

After weeks or training and unit exercises the division was moved to Avignon to defend against a possible enemy landing in the Mediterranean area.

The long-awaited alert order arrived on the morning of 10 July 1943. During the previous night Allied troops had landed on the southern tip of Sicily. On the afternoon of 11 July the units of the division boarded Ju 52 transports which, operating in several waves, flew them to airfields around Rome.

The next day *Fallsch.Jäg.R. 3* under *Oberstleutnant* Heilmann took off for Sicily on the new division's first jump mission. The paratroopers jumped south of Catania, their mission to close a gap in the front east of Carlentini and establish contact with the battle group of the Fallsch.PD. *"Hermann Göring"* deployed there.

In the next two weeks *Fallsch.Jäg.R. 4*, the machine-gun and pioneer battalions, *I. Abt.* of the artillery regiment and elements of the *Pz.Jäg. Abt.* were also dropped. By the time these forces arrived, however, the defenders were already withdrawing. The Allies were simply too strong.

One incident typified the fighting on Sicily. British commandos were dropped east of Carlentini with the objective of storming the Malati bridge. Shortly thereafter, on 13 July, paratroopers of the British 1st Brigade landed in the Plain of Catania to take the Simeto bridge. When the British succeeded in breaking through the German front from the south near Carlentini, the *Fallsch.Jäg.R. 3* was encircled.

The regiment managed to break out of the encirclement. Moving in single file, the paratroopers passed beneath the Malati bridge, over which British tanks were already rolling. Late on the evening of 16 July the paratroopers reached the German lines south of Misterbianco. At the same time British and German paratroopers clashed for the first time, at the Simeto bridge. The *Fallsch.Jäg.R. 4* (*Oberst* Walther), in particular, distinguished itself there. In the face of Allied superiority on the ground and in the air, retreat was unavoidable. The men of the *Fallsch.Jäg.R. 3* defended the fortress of Centuripe on Mount Etna to the end, making possible the German retreat to Messina. As well, paratroopers were the last German soldiers to cross the Strait of Messina to the mainland on 17 August 1943.

The *1. Fallsch.Jäg.Div.* now received orders to defend Apulia. The division's command post was located in Alta Mura. But before the paratroopers could establish themselves there, troops of the British 8th Army landed in Calabria. On 8 September the *Fallsch.Jäg.R. 4* repulsed another landing attempt near Pizzo. Italy surrendered the same day.

One day later, units of the American 5th Army with powerful air support landed on the Gulf of Salerno. The *1. Fallsch.Jäg.Div.* was immediately ordered out of Calabria. The first unit to arrive in the new combat zone, *I./Fallsch.Jäg.R. 3* (approximately 100 men), stormed and recaptured the town of Battipaglia. Its sister regiment took Altavilla after fierce fighting. Beginning on 18 September the German forces were forced to retreat; for the first time in months the *1. Fallsch.Jäg.Div.* saw action as a unit. It was pulled out of the front lines and transferred to a "quiet sector" in the Abruzzia.

In the meantime the German command had ordered the fortification of the "Gustav Line" in southern Italy. The line followed the valleys of the Garigliano and Sangro rivers and its most vulnerable area was the Liri Valley with Monte Maio and Monte Cassino. The latter would later play a fateful role in the history of the *Fallschirmjäger*. On top of the mountain was the Benedictine monastery, which was home to a valuable collection of antiquities.

During the period from 13 October to 8 December 1943 soldiers of the *1. Fallsch.Jäg.Div.* transferred these treasures from the monastery to the Vatican.

The battles of Monte Cassino were not long in starting. The Allied attack on the Liri Valley began on 17 January 1944. Shells and bombs rained down on the German positions and enemy troops broke into the defense line. At the beginning of February American, British, French and Algerian troops tried to take Mount Calvary and Monte Cassino. The *1. Fallsch.Jäg.Div.*'s hour had come.

The first battle group from the division under the command of *Oberst* Schulz (*Fallsch.Jäg.R. 1*, *III./Fallsch.Jäg.R. 3* and the *Fallsch.MG-Btl. 1*) began reaching the battlefield on 10 February 1944. It recaptured Mount Calvary and occupied the monastery hill. Not a single German soldier entered the monastery itself. (On 15 February an Allied bombing raid leveled the monastery.)

Three days later the Allies gave up the battle for Monte Cassino. The *1. Fallsch.Jäg.Div.* received orders to defend the entire sector of the Monte Cassino front.

The Second Battle of Monte Cassino began on 15 March with a powerful air attack and subsequent artillery barrage. From this point on the *Fallschirmjäger* were in the center point of a new Allied offensive. The *II./Fallsch.Jäg.R. 3* and *I./Fallsch.Jäg.R. 4* especially distinguished themselves. At the end of March the division was pulled out of the front for a week, however at the beginning of April it was sent back in. Together with the machine-gun battalion the *Fallsch.Jäg.R. 4* took over the defense of the town and the monastery hill. *Fallsch.Jäg.R. 3* was positioned in the front between Mount Calvary and Monte Cassino. By now the regiments were down to two battalions; each regiment had released its *III. Bataillon* for the formation of the new *5. Fallsch.Jäg.Div.*

The next Battle of Monte Cassino began on 11 May 1944 after a fierce artillery barrage. The Polish II Corps launched a frontal attack against the positions of the *1. Fallsch.Jäg.Div.*. It soon became apparent that the main effort was directed against Mount Calvary. The *Fallsch.Jäg.R. 3* repulsed every enemy attack and recaptured Mount Calvary. Then, without warning, on the evening of 17 May the order was received to abandon the positions.

French divisions had broken through on the other side of the German front and were driving into the rear of the Monte Cassino defenders. At the same time the American VI Corps had forced a breakout from the Anzio-Nettuno bridgehead and were now also behind the German front in the Liri Valley. The way was now open for the American divisions to advance to Rome, which they reached on 4 June 1944.

And so the paratroopers who had defended their positions so valiantly were forced to join the German retreat to central Italy. The units remained attached to the 10. Armee. In June its units occupied positions in the Rimini area and in November the men of the *1. Fallsch.Jäg.Div.* were deployed around Imola.

The *Fallschirmjäger* were now in the sector on the coast of the Adriatic Sea. Bad weather had moved in and the rain restricted movement. Both friend and foe suffered from the effects and both sides went over to positional warfare. In December it began to snow and the war ground to a halt.

The positional warfare on the Adriatic lasted until at the beginning of April 1945. Then, on the evening of 9 April, 1,500 guns of the British 8th Army opened fire. German resistance crumbled and when, on the second day of the battle, the neighboring division withdrew beyond Lake Comacchio to the Po, the *1. Fallsch.Jäg.Div.* was also forced to abandon its positions.

By this time there no longer was an intact *1. Fallsch.Jäg.Div.* Since March rumors had been circulating that the division was to be relieved. The first units had already been pulled out of the front before the British offensive started, placed on trains and transported to western Germany. The last units began leaving Italy on 12 April 1945. As they arrived, the battle groups were inserted into the Rhine defense front between Trier and Neuss. During the final days of the war they were caught up in the maelstrom of the surrender somewhere east of the Rhine.

Commanders:

Generalleutnant Heidrich; *Generalmajor* Schulz.

2. Fallschirmjägerdivision

This division, whose emblem was a comet, was formed in Normandy in February 1943. The core of the new division consisted of *Fallsch.Jäg.R. 2* and *II./Fallsch.AR. 1*. In the weeks that followed, additional units were added to bring the *2. Fallschirmjägerdivision* up to full strength:

Headquarters
Fallsch.Jäg.R. 2, *6* and *7*, each with three battalions,
Fallsch.AR. 2 with one battalion of four batteries,
Fallsch.Pz.Jäg.Abt. 2 with six companies,
Fallsch.Pi.Btl. 2 with four companies,
Fallsch.NA. 2 with two companies,
Fallsch.Nachschub-Btl. 2 (supply battalion).

Later, in the summer of 1943 with the division already in action, the *2. Fallsch.Jäg.D.* was joined by the *Fallsch.MG-Btl. 2* and the *Fallsch. Flak-Abt. 2*. These units were made up of soldiers from *Fallsch.Jäg.R. 5* (*Oberst* Barenthin) which had met its end in Tunisia.

By that time the *2. Fallschirmjägerdivision* had been transferred to southern France and was quartered in the area around Nimes. The division was held at a state of readiness there as the German command expected an Allied landing somewhere in the Mediterranean after the surrender of Army Group Tunisia.

After Italy's departure from the Axis alliance on 25 July 1943, in the south of France the *XI. Fliegerkorps* (*General der Flieger* Student), of which the *2. Fallschirmjägerdivision* (*Generalleutnant* Ramcke) was a part, was placed on alert.

On 26 July the division was transported to Rome in 90 Ju 52s, 45 He 111s and 80 gliders. Led by operations officer (Ia) General Staff *Major* Baron von der Heydte, the division's advance detachment landed at Viterbo and Practica di Mare near Rome, taking the Italians completely by surprise. The *II./Fallsch.Jäg.R. 6* took the airfields near Foggia. After landing, the paratroopers took the area around Ostia in a single operation.

The Italian commands reacted quickly. In a few days they assembled two armored divisions, one motorized division and a coastal defense division around Rome. The former allies now faced each other across the battle lines. The *2. Fallschirmjägerdivision* and the *3. Pz.Gren.Div.*, which had been brought in by *General* Student, were too weak to break through the cordon of Italian troops around Rome.

On the night of 9 September – the day of Italy's surrender – the code word "Axis" unleashed the next German operations. The II./*Fallsch.Jäg.R. 6* under *Major* Gericke took off that morning. The paratroopers jumped over Monte Retondo, the location of the Italian headquarters. After fierce fighting against Italian guard troops, in the afternoon the headquarters of the Italian armed forces fell into the paratroopers' hands.

That same morning *General* Student gave the order for the attack on the Italian capital. His two divisions moved on Rome from the south. The *2. Fallschirmjägerdivision*, organized into two battle groups under *Oberstleutnant* Meder-Eggebert and General Staff *Major* von der Heydte, advanced on either side of the Via Appia and the Via Ostiense toward Rome. During the advance the *Fallsch.Jäg.R. 6* became engaged in heavy fighting with the Italian *Piacenza* Division.

During the afternoon of the same day the Italian troops began to falter, with many of them abandoning their weapons and fleeing. The advance on Rome continued. On 10 September *Fallsch.Jäg.R. 2* reached the Stazione Termini in the heart of the city. The *Fallsch.Jäg.R. 6* arrived at the Colosseum at about the same time. The Italian military commander surrendered at 1530 hours on 10 September.

The code word "Axis" was also the signal for *General* Student to free Benito Mussolini, the Italian head of state who had been taken prisoner. After the Duce's location had been determined, on 12 September gliders carrying a battle group of the *1./Fallsch.Lehr-Btl.* (Oblt. von Berlepsch) and a special detachment of the SS landed on the Gran Sasso massif in the Abruzzia and freed Mussolini. At the same time, in the valley below the other companies of the battalion under the command of *Major* Mors seized the Gran Sasso cable car station.

In mid-September British forces occupied the Aegean islands of Leros, Samos and Cos. The German response came on 3 October. As part of the recapture of the islands, on 12 November 1943 the *I./Fallsch.Jäg.R. 2* (*Major* Kühne) landed on Leros, clearing the way for the infantry battalions coming by sea to occupy the islands.

Several days after Italy's surrender the division received entraining orders. It returned to Germany and the soon afterwards the paratroopers were on trains rolling toward Russia, already in the grip of winter. This ended the division's history as a parachute division, for from that point until the end of the war in 1945 it served exclusively as an infantry division.

The paratroopers detrained in the Zhitomir area. The division was attached to the *XXXXII. AK.* of the *4. PzArmee*. Beginning in December 1943, the *Fallschirmjäger* took part in Army Group South's fighting withdrawals. January 1944 found the division in the Krivoy Rog area and the next month the paratroopers fought in the Korsun area. There the Soviet 2nd Ukrainian Front had broken through the positions of the *8. Armee*. Elements of the *2. Fallschirmjägerdivision* were overrun by the 5th Guards Tank Army. Other elements were encircled north of Kirovograd with the *10. Pz.Gren.Div.* and the *376. ID*.

The division next participated in varied actions in the Kirovograd-Cherkassy area and the retreat through Novo Arkhangelsk east of Uman to Kishinev. There the division was attached to the XVII. AK of the 6. Armee. The last battle fought by the paratroopers in the east occurred between Balabauesti and Vainowo in Romania. Then orders were received for the relief of the battered division.

The units of the *2. Fallschirmjägerdivision* were rested and brought up to strength in the Cologne area, after which they were sent west. The first regiment to transferred to France for further training was *Fallsch. Jäg.R. 6* under *Oberstleutnant* von der Heydte. The regiment's assigned training area was Carentan in Normandy.

It was precisely there that the American 101st Airborne Division was dropped on the morning of 6 June 1944. The *Fallschirmjäger* were the first German troops to engage the Americans and there was fierce fighting throughout the day. The regiment's *I. Bataillon* under Hptm. Preikschat grimly defended St. Marie-du-Mont, the first key point of the Battle of Normandy. The paratroopers held out for five days, then, having been outflanked on the left and right, they evacuated Carentan.

The bulk of the *2. Fallschirmjägerdivision* had previously arrived in Brest on the Atlantic coast for further training. When American forces shattered the German front in Normandy and advanced deep into southern and western France, Brest became a key point in the battle for France.

In addition to the *Fallsch.Jäg.R.* 6 at Carentan, the division had also been required to send a battle group under *Oberstleutnant* Rolschewski (*III./Fallsch.Jäg.R. 7* and *3./Fallsch.AR. 2*) to St. Malo. Following the

American breakthrough and the advance in the direction of Brest by the U.S. VIII Corps (four infantry and two armored divisions), *Generalleutnant* Ramcke immediately deployed battle groups to fan out to the east. Since there were no vehicles, the paratroopers had to move on foot. These battle groups were subjected to frequent ambushes by French partisans. Casualties in the period to 12 August totaled 50 killed and 200 wounded.

On 8 August the Americans called for Brest to surrender. The commander of the fortress, *Oberst* von der Mosel, rejected the appeal. On 10 August *Generalleutnant* Ramcke was placed in command. He immediately recalled his units into the fortress. *Fallsch.Jäg.R.* 7 and the *Fallsch. Pi.Btl. 2* took over the defense sector in the eastern part of the fortress, *Fallsch.Jäg.R. 2* the western part. Command of the division was passed to *Oberst* Kroh.

In addition to his division, *Generalleutnant* Ramcke also had the 343. ID. and several navy units under his command. The German forces in Brest included 11,444 members of the army, 11,718 of the *Kriegsmarine*, 4,261 of the *Luftwaffe* and 1,990 personnel from other organizations, such as the *Organisation Todt.*

On 12 August the American air force began bombing the fortress, while American and British warships pounded the harbor area. The land battle against French partisans and the American VIII Corps began on 20 August.

There was fierce fighting everywhere, especially for possession of the fortifications on Le Couquet, where the *Fallsch.Jäg.R. 2* (*Oberstleutnant* Tannert) literally fought for every foot of ground. The *I./ Fallsch.Jäg.R. 7* under *Major* Hammer fought with equal determination at Gouesnou.

The Americans abandoned their direct assault and turned to day and night bombing of the fortress. Not until 1 September did they resume their land offensive. It took the American troops ten days to reach the outskirts of Brest. For the next nineteen days there was hand-to-hand fighting in the ruined houses and streets of Brest.

On the morning of 19 September in the wrecked buildings of the French naval school, the last German assault gun fired its last round of ammunition; the artillery and flak batteries had long since run out of ammunition. One after another, the *Fallschirmjäger* battle groups were forced to surrender. One of the last to surrender was *General der Fallschirmtruppen* Ramcke (recently promoted), who left his command post perched on a rocky cliff overlooking the sea on 20 September. With him were three officers and men of the navy radio team and the general's aides.

In recognition of his accomplishments and outstanding command of his units, *General* Ramcke was awarded the Knight's Cross with Oak Leaves, Swords and Diamonds, making him the most highly-decorated member of the German parachute corps in the Second World War.

In November-December 1944, in the Amersfoort area, the OKL formed a new *2. Fallschirmjägerdivision* from replacement units of the parachute corps. The new division lacked any sort of heavy weapons, however, and the regiments – which were assigned the same numbers as those of the original division – had only two battalions at first. At the beginning of December 1944 the new *Fallsch.Jäg.R.* 6 under *Oberstleutnant* Baron von der Heydte was assigned to carry out the last airborne mission by German paratroopers in the Second World War. The order was brief and to the point: jump over the Hohen Venn and keep the mountain roads open for the I. SS-Pz.K. at the start of the Ardennes offensive.

Two *Transportflieger* groups, with a total of 67 Ju 52s, took off at 0300 hours on 17 December 1944. Ten aircraft were shot down by enemy (and German!) flak while overflying the front, and eight other machines dropped their paratroopers in the wrong place.

In the end just 450 officers, non-commissioned officers and enlisted men landed in the target area between Malmédy and Amblève. Many came down in barely-frozen swamps, others in snow-covered forests, and only a few on roads and paths. The commander was injured on landing, but was able to assemble about 100 members of his regiment.

The small force took up positions on the road and waited for the German tanks to come. But they never did. Instead American vehicles appeared on the road and the paratroopers opened fire. When the combat vehicles "with the Iron Cross" failed to appear by evening, the *Oberstleutnant* gave the order for the paratroopers to make a break for the German lines. After a three-day march, by which time the number of paratroopers with the commander had risen to about 300, they heard the front. But the Americans were blocking the way. Only a handful of men made it through to the east. *Oberstleutnant* Baron von der Heydte and the remnants of his battle group surrendered to the Americans near Monschau. The last airborne operation by German parachute troops had come to an unfortunate end.

At the beginning of January 1945 the new division was transferred to the front. It was attached to the *LXXXVIII. AK.* of the *25. Armee* in the Arnhem combat zone. In February the division was forced to join the general retreat. Given the enormous superiority enjoyed by the attacking British and Canadian forces, especially in armor and air power, the *Fallschirmjäger* fought with great determination, never voluntarily giving up any ground. The paratroopers fought in and around Gennep, Üdem, Grootehorst, Siebengewald, Baal and, in March, Rheinsberg and Xanten. There the division crossed the Rhine to Mühlheim an der Ruhr. There the *LXIII. AK.* with the *2. Fallschirmjägerdivision* was split off from its army group by the Americans. They were subsequently attached to Army Group B commanded by *Feldmarschall* Model, who had declared the Ruhr region to be a "fortress" after the American 3rd and 9th Armies encircled the army group at the beginning of April. A report from those days:

> "Battle groups of the *2. Fallschirmjägerdivision* under *Generalleutnant* Lackner resisted in and around Castrop-Rauxel. They fought in Habinghorst on 4 April, the next day in the northern part of the city, especially at the Zechen line, and one day later at the Rauxel station. 7 April brought fighting at the Emschertal line and in the rural districts of Deininghausen, Dingen, Schwerin and Frohline. The battle for Castrop-Rauxel ended on 8 April when the old city of Castrop, plus Upper Castrop, Behringhausen and Merklinde fell. Worn down by continuous artillery fire, tank fire and air attacks, the paratroopers and the Volkssturm people attached to them gave up the struggle."

The history of the *2. Fallschirmjägerdivision* had ended for a second time.

Commanders:

Generalleutnant Ramcke; *Generalmajor* Kroh; *Generalleutnant* Lackner.

3. Fallschirmjägerdivision

As a result of the fighting in Italy and the lessons learned there, the OKL formed two new parachute divisions. The *3. Fallschirmjägerdivision* was formed at troop training grounds and barracks camps in and around Reims in the autumn of 1943. Its units were made up of volunteers from the parachute training units and the former *Luftwaffe* field divisions. Placed in command of the division was *Generalmajor* Heilmann, one of the original German paratroopers.

The new division was organized as follows:

Headquarters with headquarters company,
Fallsch.Jäg.R. 5, *8* and *9*, each with three battalions,
I./Fallsch.AR. 3,
Fallsch.Pz.Jäg.Abt. 3,
Fallsch.Pi.Btl. 3 with four companies,
Fallsch.NA. 3 with two companies,
Fallsch.Versorgungstruppen 3 (supply units).

In April 1944 the division transferred to the area around Brest in western France. By this time the division was commanded by *Generalleutnant* Schimpf, who was to lead the unit in its first actions.

The alert order was received on 7 June, one day before the Allied invasion of France. The division was rushed to the Normandy front and on 8 June it was near St. Amand. This moved required the units to complete a foot march of approximately 250 kilometers and on arrival they were sent into the battlefield at St. Lô without rest. The *Fallschirmjäger* participated in the battles at St. Lô, which was the focal point of the fighting in the American sector in the first weeks after the invasion. Along with the *17. SS-Pz.Gren.Div*, the *3. Fallschirmjägerdivision* held the sector of front in the Vire Valley.

The division was on the left wing of the *II. Fallsch.K.* when the Americans succeeded in breaching the German front near Avranches and launching a drive deep into central France. The parachute troops were forced to turn their front to the west and then the southwest. The American armored forces had meanwhile advanced farther to the south and southwest, and when they occupied Le Mans on 8 August they closed the ring around the German *7. Armee*.

Contrary to Hitler's order to "hold at all costs", the *Armee*'s commander-in-chief, *SS-Obergruppenführer* Hausser, instructed it to break out of the Falaise Pocket. The *3. Fallschirmjägerdivision* was northwest of Flers, its front facing west. When the order to breakout was received, the *II. Fallsch.K.* formed three battle groups which were to smash a hole in the Allied front between Chambois and Argentan.

With the support of SS units, the *Fallsch.Jäg.R. 9* (*Major* Stephani) and *Fallsch.Jäg.R. 15* (of the *5. Fallsch. Jäg.D.*) were able to break through the Anglo-American positions. *Generalleutnant* Schimpf, who was wounded in this action, immediately dispatched all of his units toward the gap. The commanding general of the *II. Fallsch.K.*, *General der Fallschirmtruppen* Meindl had the Commander-in-Chief of the *7. Armee* brought behind the parachute troops, saving him.

The entire *II. Fallsch.K.* was transferred to the Cologne area to rest and refit and by September it was ready for action again as part of the newly-formed *1. Fallsch.Armee*. The army had received orders to defend the area between Antwerp and the Albert Canal in Belgium.

It was only a few days before the division saw action again. On 17 September 1944 the Allies launched "Operation Market Garden" between Arnhem and Nijmwegen. Two British and one American airborne divisions jumped that day, their objectives to secure and hold the bridges over the Maas, Waal and Lower Rhine for an advance into northern Germany. The alarm was sounded and the *3. Fallschirmjägerdivision* crossed the Rhine north of Emmerich. The division successfully defended the area between the Rhine and the Waal against all enemy attacks until the end of September.

A battle group from the division also took part in an offensive action. At the beginning of September it was attached to the 5. Pz Armee. The battle group, which also included units of the *5. Fallsch.D.*, a panzer-grenadier regiment and two infantry regiments, was supposed to close a gap in the front between Nancy and Lunéville. The battle group was led by Hptm. Milch, commander of the *Fallsch.Granatwerfer-Versuchs-Btl.* (parachute mortar test battalion).

During the course of the fighting withdrawal to the borders of the Reich, in the period beginning October 1944 the division fought in the Mreode and Jüngersdorf area in the Düren area on the Lower Rhine. Then a new transfer order was received.

Preparations for "Operation Watch on the Rhine", the Ardennes offensive, had begun.

The *3. Fallschirmjägerdivision*, now under the command of *Generalmajor* Wadehn, and the *5. Fallschirmjägerdivision* were the only major units of the parachute corps to take part in the offensive. The division was positioned on the right wing of the *6. SS-Pz.Armee*. The latter's units were part of the main effort with a long-range objective of Antwerp. The paratroopers were attached to the I. SS-Pz.K. At the beginning of December the division moved into its assembly area southeast of Elsenborn in heavy blowing snow.

At 0530 hours on 16 December 1944 2,000 guns opened the *Wehrmacht*'s last major offensive. When the paratroopers moved out of their positions in the cold and damp winter weather, they saw that the barrage had inflicted little damage on the American trenches. Within minutes the enemy's artillery opened fire and inflicted the first casualties.

The second day of the attack saw the paratroopers in the Lanzerath area, south of the large troop training grounds at Elsenborn. There the advance stopped. However, since the *3. Fallschirmjägerdivision* had advanced farther than any other infantry division, the *SS-Pz.Rgt. 1* committed its latest Tiger tanks there and temporarily created breathing space for the paratroopers. Then, on the third day of the attack, the 12. SS-Pz.Div. was inserted to the right of the *3. Fallschirmjägerdivision* and attacked in the direction of Krinkelt. As a result the parachute regiments were turned to the northwest to prevent any gaps from forming in the German line. Losses were terribly high.

On Christmas Eve 1944 the *3. Fallschirmjägerdivision* was ten kilometers south of Malmedy with its front facing northwest in combat against the American 30th Division. Then the division's strength was exhausted.

The retreat began in January 1945 and the division withdrew through the Eifel region into the area east of Bonn. Later the division was surrounded with the rest of Army Group B in the so-called "Ruhr Pocket".

The division's last radio message to the army command was made on 15 April 1945 from the area around Marienheide:

"We are signing off!"

Commanders:

Generalleutnant Schimpf; *Generalmajor* Wadehn; *Generalleutnant* Schimpf.

The first soldiers of the parachute corps, which evolved from the "General Göring" Regiment, marching through Stendal. After packing and jump training, parachute troops walk out to Ju 52 transports in preparation for their first jump.

The first jump: after the jump master gave the order "Ready to jump!" the paratrooper stepped into the open doorway of the aircraft. When the klaxon sounded, the parachutist jumped.

A stick of 36 parachute troops jumping from three Ju 52 transports.

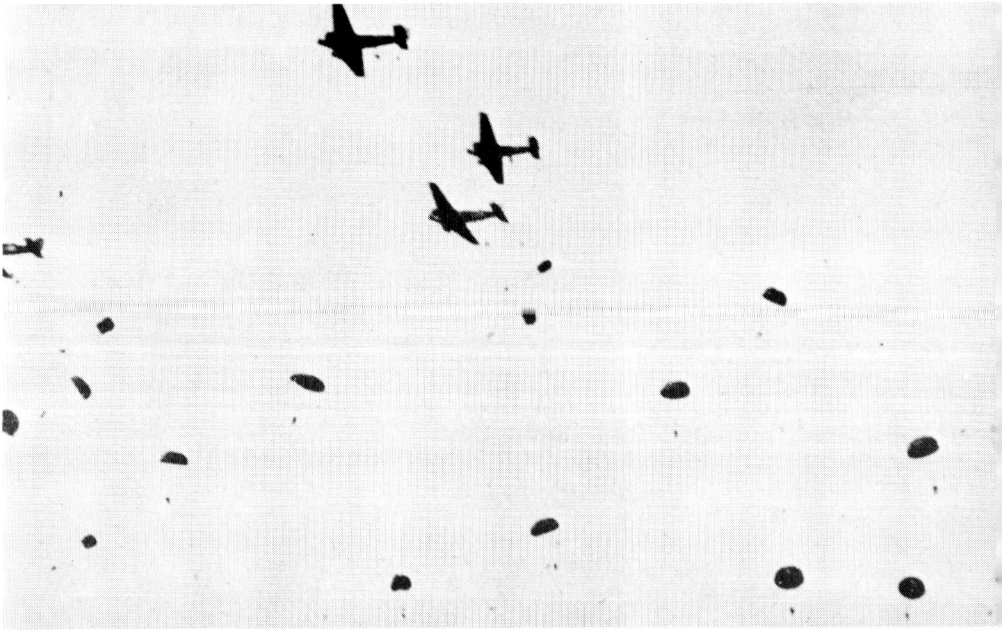

General der Fallschirmtruppe Student, commanding general of the XI. Fliegerkorps, speaks to a paratrooper after his first jump. Next to Student is his operations officer, General Staff Major von Roon.

The first public appearance by the new parachute corps took place on 20 April 1939 during a parade in Berlin to mark Adolf Hitler's fiftieth birthday.

The German parachute corps began preparing for the western campaign in November 1939, when "Sturmabteilung Koch" was formed in Hildesheim. The unit's mission was to seize the Belgian fortifications and the big bridges over the rivers and canals in the Dutch-Belgian border region. Troop-carrying gliders were used for the first time. Illustrated here is the DFS 230, which could accommodate ten paratroopers.

One of the objectives was the 1 200-meter-long road bridge near Moerdijk, which was taken by II./ Fallsch.Jäg.R. 1 on the morning of 10 May 1940.

Six members of "Sturmabteilung Koch", which took the Belgian frontier fort of Eben Emael after fierce fighting.

The parachute troops which were sent to take Dutch airfields and attack Rotterdam met fierce resistance and made little progress for several days.

Heavy ground fire made it almost impossible for the Ju 52s to land at Ypenburg near The Hague, and as a result ten aircraft came down on the highway between The Hague and Rotterdam. The parachute troops were finally relieved when leading elements of the 9. Pz.D. arrived at the Dordrecht bridge. The fourth officer from the left is Oberst Bräuer, one of the pioneers of the parachute corps.

The year 1941 brought an expansion of the war into the Balkans. Fallsch.Jäg.R. 2 received orders to storm the bridge over the Corinth Canal. Here paratroopers race across the bridge under British fire on 26 April 1941. This was the last photo taken by a well-known photographer who was killed in this action.

The parachute corps' most important action, and its costliest, was the airborne invasion of Crete. On the morning of 22 May 1941 paratroopers were dropped at three locations simultaneously, from Maleme in the west to Iraklion in the east. The targets were airfields and the capital of Chanea. Some of the parachute troops came down in the positions of the defenders, who had been alerted and were waiting for the attack. Enemy resistance was not broken until the 5. Geb.D. came ashore.

Parachute troops open one of the air-dropped weapons canisters and remove the light machine-guns and rifles inside. They then form up into squads and attack the heavily-defended British and New Zealand positions.

The struggle for the cities, towns and hills of Crete went on for weeks, before the enemy was forced back to the south coast at the end of May. These photos show the capture of the last New Zealand defenders in the area of Sfakia. The German parachute corps lost 3,250 men killed and 3,400 wounded in the battle.

When the German offensive against the Soviet Union ground to a halt in the winter of 1941, parachute troops were committed as infantry in the front lines. The photo above depicts the much-fought-over airfield at Zhaykovka, which I./Fallsch.Sturm.R. defended at heavy cost. Below: 2. Fallsch.Jäg.D. on the retreat in the Ukraine in 1944.

When the German Africa Corps' advance bogged down at El Alamein on the Egyptian frontier in 1942, the Fallsch.Brigade Ramcke was dispatched to North Africa. The soldiers, with almost no vehicles, made their way back to Tunisia through the desert sands. Above paratroopers manhandle an anti-tank gun through the desert, below an unserviceable motorcycle-sidecar.

The next theater of war for the German parachute corps was Italy. Fallsch.Jäg.R. 3 parachuted into Sicily, fought its way through enemy lines south to Catania and was received by the division commander. Following the departure of Italy from the war, on 9 September 1943 the 2. Fallsch.Jäg.D. stormed Rome (below).

The year 1944 saw the parachute troops in action in Italy, where they earned the name "Green Devils" in the Battle of Monte Cassino. The photo above shows the monastery on Monte Cassino after it was destroyed by Allied bombing on 15 February 1944. German parachute troops held out for months, contesting every meter of ground.

The last major action by the German parachute corps took place in France in 1944. The parachute troops were in action from the very first day of the Allied invasion and remained effective until the end of the war. Below a strongpoint on the invasion front with MG 42 machine-gun.

General der Fallschirmtruppe Ramcke.

4. Fallschirmjägerdivision

This division was formed at the same time as the *3. Fallschirmjägerdivision*. Formation took place in the Perugia area. The new unit was commanded by *Generalmajor* Trettner – *General* Student's 1st General Staff Officer and Inspector-General of the Bundeswehr after the war. Composition of the regiments and division units included elements of the 2. Fallsch.Jäg.D. which had not been transferred to the Eastern Front, volunteers from the disbanded Italian Dembo and Folgore divisions and personnel sent from Germany.

The division consisted of:

Headquarters with headquarters company,
Fallsch.Jäg.R. 10, *11* and *12*, each with three battalions,
I./Fallsch.AR. 4,
Fallsch.Pz.Jäg.Abt. 4 with three companies,
Fallsch.Pi.Btl 4 with four companies,
Fallsch.NA. 4 with two companies,
Fallsch.Versorgungstruppen 4.

The *4. Fallschirmjägerdivision* was to remain in the Italian theater until the end of the war and took part in every decisive battle there from the beginning of 1944.

The division's first action took place at the end of January when the paratroopers were still in training. With covering fire from numerous warships, American and British troops landed in the Bay of Anzio on the morning of 22 January 1944. Surprise was total: there were no German defenders north of Anzio. All nearby German units – and there were not many – were immediately placed on alert.

Among them was an assault gun battery of the *4. Fallschirmjägerdivision* which happened to be on the march on the Albano – Nettuno road. The first shells were fired at 0715 hours. The towns of Anzio and Nettuno were firmly in Allied hands, however, and one hour later the two beachheads were linked up.

The *4. Fallschirmjägerdivision* was immediately ordered to move out, as was the Fallsch.PD. "Hermann Göring", which was also in training. The first unit to reach the battlefield was a battalion of the *Fallsch.Jäg.R.* 11 under *Major* Gericke, a veteran of Norway. By evening additional batteries and platoons of the army and air force had also reached the battlefield; but by this time the Allies already had 36,000 men and 3,200 vehicles ashore.

On 25 January the corps headquarters of the I. Fallsch.K. (*General der Flieger* Schlemm) assumed command of the German forces around the Allied bridgehead. He had available three divisions: the Fallsch.Pz.Div. "Hermann Göring" on the coast near Terracino and Carano, the 3. Pz.Gren.Div. on either side of the Albano road and the *4. Fallschirmjägerdivision* to the mouth of the Tiber. The supply and fuel dumps were located in the rear and were subjected to constant bombing attacks which destroyed or damaged them.

In the days that followed, bad weather hampered the fighting. Nevertheless, the German command succeeded in moving additional units to the front by 30 January. This allowed the 4. Fallsch.Jäg.D. to shorten its previously extended front. On that day an attack was launched by the American VI Corps, however it was halted near Cisterna. A German counterattack launched on 3 February 1944 was not a success. The division was forced to return to its starting position. Then it began to rain again. Another attack was made and German troops held out in Aprilia for several hours; another attack on Carroceto by the *Fallsch.Jäg.R. 11* and *GR. 147* also failed.

The second part of the German counterattack began in clear weather on 14 February. The paratroopers played no part, instead securing right flank of the attack. The attack was smashed by Allied warplanes, artillery and tanks. At the end of February *Feldmarschall* Kesselring called off further attacks.

The front around the Anzio-Nettuno bridgehead was relatively quiet in March and April; friend and foe needed a pause to catch their breath. The next Allied offensive from the bridgehead began on 22 May 1944. The battle lasted four days, then the German front collapsed. The *Fallsch.Jäg.Sturm-R.* (*Major* Timm) was surrounded but held out. By the end of May the battle was lost; the Allies had opened the road to Rome.

On 1 June the *4. Fallschirmjägerdivision* was on the extreme right flank of the German front south of Rome. The paratroopers defended a line between the coast and Valmontone. Three days later battle groups of the division marched through Rome's eastern suburbs across the Aniene. German troops abandoned the Italian capital. The division had suffered heavy casualties in the fighting to date. For example, the *4. Fallschirmjägerdivision*'s combat strength on 1 July 1944 was 140 officers, 831 non-commissioned officers and 3,083 enlisted men.

The subsequent retreat took the paratroopers into the Florence area. By the end of July the division lost 1,008 men killed, wounded or missing. At the beginning of August troops of the American 5th Army attacked the city. Combat engineers of the Fallsch.Jäg.Pi.Btl. 4 placed explosive charges on the bridges, however they could not be detonated. The retreat to the north began, however the onset of winter weather prevented the Allies from pursuing vigorously.

The I. Fallsch.K., including the 1. and 4. Fallsch.Jäg.D., remained in the Adriatic sector. Fighting stagnated there in the first three months of 1945. Not until 9 April 1945 did the British 8th Army resume the offensive against the German 10. Armee, of which the paratroopers were a part. On the second day the defenders were forced to retreat beyond Lake Comacchio in the direction of the Po. One day later a gap appeared between Bologna and the Reno River. The British advanced into this gap toward Bondeno west of Ferrara, splitting the two German armies.

It was the start of the final battle in Italy. By the end of April 1945 the *4. Fallschirmjägerdivision* consisted of isolated battle groups in positions between Trient and Lake Comer. The division's history ended there with the surrender of the German forces.

Commander:
Generalleutnant Trettner.

5. Fallschirmjägerdivision

In the spring of 1944 the losses suffered by the German parachute corps in the fighting for Sicily and the Battles of Monte Cassino led the OKL to form a fifth parachute division. Units of the first three divisions, soldiers from the air training and replacement regiments and members of disbanded *Luftwaffe* field divisions and bomber units were transferred to the Reims area, where they were combined into the *5. Fallschirmjägerdivision*.

Placed in command of the division was *Generalleutnant* Wilke, formerly commander of the 1. Luftw.Feld-D. in the Novgorod area on Lake Ilmen (northern Russia). Under his command were:

Fallsch.Jäg.R. 13, *14* and *15*, each with three battalions,
I./Fallsch.AR. 5,
Fallsch.Pz.Jäg.Abt. 5 with four companies,
Fallsch.Pi.Btl. 5,
Fallsch.NA. 5,
Fallsch.Nachschubtr. 5.

The new division was first stationed at troop training grounds and barracks camps around Reims, Chateau-Thierry and Châlons sur Marne. In May 1944 it was transferred into the Rennes area in Brittany for further combat training. There was not much time left, for on 6 June 1944 the Allies launched their invasion of Normandy.

Following the successful enemy landings, the division received alert and combat orders. The division was rushed in trucks and on foot to the combat zone between St. Lô and Aire on the Vire. As part of the LXXXIV. AK., which also included three infantry divisions, it was able to halt the first American armored attacks.

In the weeks that followed, Allied barrage fire and continuous bombing attacks destroyed almost all of the division's heavy weapons and the makeshift positions manned by the regiments. When the Americans attacked on 26 July the division's main line of resistance collapsed. On the first day of fighting American tanks reached the St. Lô – Coutances road.

American forces shattered the German front at Avranches, drove into western and southern France and encircled the *7. Armee* in the Falaise pocket. Among the units trapped there was the *5. Fallschirmjägerdivision*. Without heavy weapons, the paratroopers lay in the woods east of Domfront and defended the wooded terrain to the south. Several battered groups of paratroopers were among the German forces to escape the pocket in the breakout ordered by the AOK 7. The division had ceased to exist as a cohesive division.

The *5. Fallsch.Jäg.D.* was pulled out of the front and brought up to strength in men and equipment in the rear. At the beginning of September 1944 it was placed under the command of the newly-formed *1. Fallsch.Armee*. A battle group from the *3.* and *5. Fallsch.Jäg.D.* under the command of *Oberst* Eggersh was deployed for a time between Nancy and Lunéville.

In November 1944 the *5. Fallschirmjägerdivision* was back at full strength and ready to return to action. Preparations for the Ardennes offensive had begun. The division was attached to the LXXXV. AK. of the 7. Armee and positioned on the army's right wing on the east bank of the Our near where it flows into the Sauer.

The 7. Armee's mission: secure the southern wing of the German attack force, whose long-range objective was Antwerp. The 7. Armee was the weakest of the German formations. The AOK had available just three corps commands with three infantry divisions, the *5. Fallschirmjägerdivision* and several security battalions.

The German offensive began on 16 December 1944 in cold and wet winter weather.

Together with the attached *Fallsch.Sturmgeschütz-Brig. 9*, the *5. Fallsch.Jäg.D.*, now under the command of *Generalmajor* Heilmann, was first supposed to cross the Clerf river line. Thanks to the tireless efforts of the assault guns this was achieved in the first hours. That evening the paratroopers entered Vianden, placing them far ahead of the rest of the *Armee*.

In spite of stiffening American resistance, in the days that followed the division was able to enter Wiltz and on 24 December was south of Bastogne, just a few kilometers north of the Sauer and near the important town of Martelange. Resistance by the American III Corps and its 26th Division halted the advance, however. The division was forced onto the defensive far ahead of the rest of the *7. Armee*, approximately 40 kilometers west of its departure positions.

The German offensive in the Ardennes had reached and passed its zenith. With clearing weather the Allied air forces struck. The German forces were unable to make any further gains. At the end of December the order was issued to fall back.

The 5. Fallsch.Jäg.D. began withdrawing from the snow-covered area south of Bastogne to its departure positions. The enemy immediately pursued energetically with tanks. In January the units of the division were in the Wiltz area; one month later the exhausted battalions were positioned around Prüm in the Eifel region. From there only a few battle groups, with no heavy weapons, reached the area of Lake Laacher. By the beginning of April 1945 the *5. Fallschirmjägerdivision* had ceased to exist as a cohesive formation. Only remnants of the division reached the Ruhr pocket.

Commanders:
Generalleutnant Wilke; *Generalmajor* Heilmann.

6. Fallschirmjägerdivision

When the *6. Fallschirmjägerdivision* was formed around Amiens, Allied bombers were already dropping thousands of tons of bombs on northern France and enemy warships were pounding the coastline of Normandy. Officers, non-commissioned officers and enlisted men were drawn from air training and air replacement regiments in France, Belgium and Holland. These were supplemented by convalescent paratroopers from home and replacement personnel from the parachute schools.

The *Fallsch.Jäg.Rgter 16*, *17* and *18* plus all the divisional units (artillery, anti-tank, combat engineer, flak, signals and supply) with the number 6 were formed while the Battle of Normandy was still raging. Placed in command of the new division was *Generalleutnant* Plocher, formerly chief-of-staff of *Luftflotte 3* in France.

While in the midst of training, the *Fallsch.Jäg.R. 16* (partly formed from the former *Fallsch.Jäg.R. 5*) was ordered to proceed immediately to Stendal. There the paratroopers were loaded aboard Ju 52 transports and flown into the surrounded fortress of Vilna. Soldiers of the army and *Waffen-SS* had been defending the Lithuanian capital since 2 July 1944.

The Ju 52's carrying *Oberstleutnant* Schirmer's *Fallsch.Jäg.R. 16* landed on Vilna's battered airfield. The Soviets attacked on 8 July and after three days of fighting the fortress commander ordered his forces to break out. What was left of the garrison, including a small remnant of the parachute regiment, forded the Vilna during the night of 13 July and in the morning reached the forward positions of the *Fallsch.Pz.K. "Hermann Göring"*.

After completing its training, the main body of the *Fallsch.Jäg.D. 6* still in France, now just two regiments strong, was attached to the *1. Fallsch.Armee* and inserted into the defense line between Antwerp and the Albert Canal. No sooner had the troops occupied their positions on the Maas when, on 17 September 1944, a bright, sunny Sunday, the Allied airborne operation between Arnhem and Nijmwegen began. Two days later the British XXX Corps attempted to relieve the landing force, which had been encircled by troops of *Feldmarschall* Model. British tanks attacked at about 0230 hours on 18 September, splitting the *85. ID* and the *6. Fallsch.Jäg.D.* The British advanced as far as Eindhoven and established contact with the American 101st Airborne Division.

The paratroopers held together, however, and for the next ten days they fought 20 kilometers southeast of Eindhoven and far west of the Maas against far superior formations of the British 2nd Army. The division held its positions until 28 September 1944, when the Battle of Arnhem ended as a ringing defensive success for the German side. The paratroopers were still in the Maas position when 1945 began.

In the final months of the war the division was attached to the II. Fallsch.K. and took part in the corps' retreat which began in February 1945. The battalions and companies fought a number of bitter engagements against the pursuing British and Canadians, especially at Xanten, but then were forced to retreat beyond the Rhine. Battle groups of the division reached Oldenburg in April 1945 shortly before the end of the war.

Commander:
Generalleutnant Plocher.

7. Fallschirmjägerdivision

In consultation with the high command of the *1. Fallsch.Armee*, on 9 October 1944 the OKL ordered the formation of a *7. Fallsch.Jäg.D.* in order to bolster the forces of the *1. Fallsch.Armee* in the northwest of the Reich and in Holland.

Battalions from the *6. Fallsch.Jäg.D.* which had been battered in the fighting at Eindhoven in September were used in the formation of the new division. They were joined by training units from the homeland as well as complete alert battalions from the army and air force. The latter came from bases in Menzel, Grossmehl, Greve, Grunewald, among others.

Generalleutnant Erdmann assumed command of the division. Under his command were the *Fallsch.Jäg.Rter 19*, 20 and 20 plus artillery, antitank, flak, signals and medical battalions and supply units with the number 7. While the division was still in the formation process, the first of its units were sent across the Maas to bolster the 6. Fallsch.Jäg.D. stationed there.

Suddenly, at the end of 1944, came the order to transfer into the area of the 1. *Armee* in Lorraine. "Operation North Wind" was supposed to lead to a German success in the Vosges and prevent the British and Americans from attacking toward the Lower Rhine.

Together with one panzer division, one panzer-grenadier division and one infantry division, the *7. Fallschirmjägerdivision* was subordinated to the XXXIX. Pz.K., which attacked toward the northern Vosges on 8 January 1945. By 14 January the corps, with the *7. Fallschirmjägerdivision* on the left wing, had advanced as far as the northern edge of the Hagenauer Forest. The paratroopers entered Rittershofen. But then the attack was halted on direct orders from Hitler and the XXXIX. Pz.K. was disbanded.

The *7. Fallsch.Jäg.D.* returned to the Lower Rhine under the command of the *II. Fallsch.K.* of the *1. Fallsch.Armee*. There the *Armee* was engaged in battle against British and Canadian forces.

For a time the division was attached to the *LXXXVI. AK.* between Emmerich and Roermund, after which it was part of the *II. Fallsch.K.* until the end of the war. The battle for the Lower Rhine began on the morning of 8 February 1945 with a massive artillery and air bombardment. The next day the division was forced to retreat toward the Reichswald. The division defended Kleve, fighting for hours against British and Canadian tanks on the road from Kalkar to Goch. Not until 16 February did the paratroopers surrender Goch. A report from the last days of February 1945 states:

> "The British XXX Corps fought its way from Goch toward Weeze, Kevelaer and Geldern. The battle for Weeze and the small farms which had been turned into strongpoints went on for days. The men of the 7. Fallsch.Jäg.D. stood firm."

The division remained part of the II. Fallsch.K. together with the 6. and 8. Fallsch.Jäg.D., a panzer division and two infantry divisions in the Wesel bridgehead. The Allied attack against this last German bridgehead on the west bank of the Rhine began at the beginning of March. Wesel had ceased to exist as a city, having been reduced to rubble.

The final battle began on 5 March. The *Fallsch.Jäg.R. 21* fought to the last man at the Grünthal crossroads near Wesel. The *3./Fallsch.Jäg.R. 20* defended Lintfort near Rheinberg for hours. Three days later the bridgehead was evacuated. The remains of the division were inserted into the front near Emmerich.

Not until the end of March were the positions on the Lower Rhine finally surrendered. At the beginning of April the 1st Canadian, 2nd British and 9th American Armies crossed the Lower Rhine. The *II. Fallsch.K.* (*6.*, *7.* and *8. Fallsch.Jäg.D.*) was forced to withdraw to Oldenburg.

The story of the *7. Fallschirmjägerdivision* ended in the final days of April 1945 between Oldenburg and Bremen.

Commander:
Generalleutnant Erdmann.

8. Fallschirmjägerdivision

In the first weeks of 1945 the OKL began combing the *Luftwaffe*'s training and replacement units in the homeland in an effort to free up all surplus officers, non-commissioned officers and enlisted men for the parachute corps. Since the existing *Luftwaffe* field divisions were under the command of the army, it was intended that at least those air force soldiers in the barracks, airfields and training fields should be placed under the command of the *Luftwaffe*. Thus the order for the creation of the *8. Fallschirmjägerdivision* was issued at the beginning of February 1945. First commander of the division was *Generalleutnant* Wadehn, former commander of the 10. Luft.Feld-D. in northern Russia.

The division consisted of the *Fallsch.Jäg.Rter 22*, *23* and *24*, although these units never achieved the strengths of normal regiments, plus all the usual division units (artillery, anti-tank, combat engineer, etc.) with the number 8. None of these units reached the strengths of the units of earlier divisions.

While the new division was still in training it was called upon to send battle groups to the Wesel bridgehead, where they were inserted into the front lines. Other units eventually arrived and the division saw action in the Bönninghard area against superior Canadian and British forces. The *Fallsch.Jäg.R. 24* distinguished itself by closing a gap in the front between Alpen and Veen. After evacuating the bridgehead at the beginning of March 1945 the division took part in defensive fighting north of Duisburg.

Until the end of March the division withdrew into the Bremen area, constantly battling British armored forces. At the beginning of April the area west of the port city was the battle zone. There the *15. Pz.Gren.D.*, the *Pz.Ers.Brig. "Grossdeutschland"* and the *8. Fallsch.Jäg.D.* were engaged in battle against troops of the Canadian II Corps and the British XXX Corps. The paratroopers defended the city of Menslage for two days, while other elements of the division fought in Schwagstorf. On 13 April the last battle groups from the division were holding out around Cloppenburg. The *8. Fallsch.Jäg.D.*'s front finally collapsed when 300 British armored vehicles moved on Cloppenburg. Remnants of the division were able to fight their way through to Lauenburg on the Elbe, where the end was unavoidable.

Commander:
Generalleutnant Wahden.

9. Fallschirmjägerdivision

Formation of the division was planned for October 1944 but did not take place until January 1945. The OKL had made *Wehrkreis II* responsible for the raising of this division. The division provided officers and men from air units which had been disbanded because of the fuel shortage, while *Wehrkreis II* added replacement units from garrisons in Pomerania. The division was put together in the rear area of the *11. Armee* and later of the *3. Pz.Armee* in Pomerania.

General der Fallschirmtruppen Bräuer, one of the parachute corps' first officers and most recently commander of the Greek islands, was named the division's first commander. He was soon replaced, however.

The division was supposed to include the *Fallsch.Jäg.Rter 25, 26* and *27*, however the formation process was never fully completed. On the other hand a Pz.Zerstörer-Abt. (tank destroyer battalion) of three companies under *Oberst* Hermann was formed. Also formed were a combat engineer battalion (three companies) and a signals battalion (two companies). The *Fallsch.AR 9* was assigned two battalions from existing *Luftwaffe* units.

The *I./Fallsch.Jäg.R.* 25 soon left the division, becoming the independent *Fallsch.Jäg.Btl. 67*. At the end of February 1945 the battalion was loaded aboard Ju 52 transports and cargo gliders and deposited around Breslau. The *II.* and *III./Fallsch.Jäg.R. 26* soon followed. These three battalions joined the garrison of Breslau, which fought on until the surrender. The *9. Fallsch.Jäg.D.* was no longer a major unit.

The division group, now under the command of *Oberst* Hermann, fought on the Oder front as part of the XI. SS-K. and at the end of April withdrew to Berlin.

The division was assigned to defense sectors G and H in the northern part of Berlin and in the days that followed defended the northeastern suburbs of the city against the onrushing tank forces of the Red Army. Since the division was still relatively intact it was selected to lead the planned breakout to the west until this was vetoed by Hitler. And so the paratroopers stayed in Berlin until the military commander surrendered the city on 2 May 1945.

Commanders:
General der Fallschirmtruppen Bräuer; *Oberst* Hermann.

10. Fallschirmjägerdivision

In March 1945, with the war obviously approaching its end, the OKL ordered the formation of a tenth parachute division, even though there were no forces in the Reich which could be considered *Fallschirmjäger.*

To provide the necessary personnel, units of the 1. and 4. Fallsch.Jäg.D. being transported from Italy to the Western Front were taken from their trains in Melk. Each of the two divisions was required to provide three complete battalions and an artillery battalion. This never took place, however. Instead of the hoped for *Fallsch.Jäg.Rter 28, 29* and *30* only the *Fallsch.Jäg.R. 30* was formed in its entirety. It was joined by the *I./Fallsch.AR. 10* and the *Fallsch.Pz.Jäg.Abt. 10* with five companies.

Oberst von Hoffmann assumed command of this "rump division", which at the end of April was transported to the *XXIV. Pz.K.* of the *1. Pz.Armee* in Moravia. There the *Fallschirmjäger* were soon engaged with Soviet armored forces rolling toward Brünn. Part of a battle group commanded by *Oberst* von Hoffmann was redirected via St. Pölten into the Alpine region where it saw action under the command of the *6. SS.Pz.Armee.*

The various battle groups of the last front-line German parachute division were swallowed up in the maelstrom of the surrender.

Commander:
Oberst von Hoffmann.

11. Fallschirmjägerdivision

In the final days of March 1945 *Luftgau XI* (Oldenburg) received orders to form an eleventh parachute division which was to include the *Rgter 37*, *38* and *39* plus division units with the number 11. A cadre of officers and non-commissioned officers was supposed to come from the *Luftwaffe*'s replacement units and from convalescent companies. The division's enlisted men were to come from the army, navy and the Volkssturm.

Even as the division was in its initial establishment phase it was put aboard trains and transported to the Netherlands. The few units were assigned to various *Luftwaffe* training fields, however they were never assembled into a cohesive combat unit and front-line use was not possible. When the surrender came the core of the division was still in training near the front.

The *11. Fallschirmjägerdivision* had no division command and therefore no commander and staff.

Appendices

Organization and Command Staff of the Parachute Corps during the Landing on Crete, 20 May 1941

Generalkommando XI. Fl.K. (Headquarters, XI Air Corps)

Commanding General:	*General der Flieger* Student
Chief-of-Staff:	*Generalmajor* Schlemm
Ia (Ops. Officer):	General Staff *Oberstleutnant* Trettner
IIa (Adjutant):	*Oberst* von Fichte
7. Fl.D. (7th Air Division):	
Division Commander:	*Generalleutnant* Süßmann (killed en route to Crete)
Ia (Ops. Officer):	General Staff *Major* Count von Üxküll
Fallsch.Jäg.R. 1:	*Oberst* Bräuer
I. Btl.:	*Major* Walther
II. Btl.:	*Hauptmann* Burckhardt
III. Btl.:	*Major* Schulz
Fallsch.Jäg.R. 2:	*Oberst* Sturm
I. Btl.:	*Major* Kroh
II. Btl.:	*Hauptmann* Pietzonka
III. Btl.:	*Hauptmann* Wiedemann
Fallsch.Jäg.R. 3:	*Oberst* Heidrich
I. Btl.:	*Hauptmann* Baron von der Heydte
II. Btl.:	*Major* Derpa
III. Btl.:	*Major* Heilmann
Luftlande-Sturm-R. (Airborne Assault Regiment):	*Generalmajor* Meindl
I. Btl.:	*Major* Koch
II. Btl.:	*Major* Stentzler
III. Btl.:	*Major* Scherber
IV. Btl.:	*Hauptmann* Gericke
Fallsch.Pi.Btl. 7:	*Major* Liebach
Fallsch.Art.Abt. 7:	*Major* Bode
Fallsch.MG-Btl. 7:	*Hauptmann* Schulz
Fallsch.Pz.Jäg.Abt. 7:	*Hauptmann* Schmitz
Fallsch.Fla-MG-Btl. XI:	*Hauptmann* Baier

Fallsch.San.Abt. XI: *Ob.Stabsarzt* Dr. Berg
Luft-NA. 41 (mot.): *Hauptmann* Hoppe

Soldiers Awarded High-Ranking Decorations

Sequential Number	Rank/Name/Post	Date
Knight's Cross with Swords and Diamonds		
20	*General der Fallschirmtruppen* Ramcke, C.O. Fortress Brest	19/9/1944
Knight's Cross with Oak Leaves and Swords		
55	*Generalleutnant* Heidrich, C.O. 1. Fallsch.Jäg.D.	25/3/1944
67	*Oberst* Heilmann, C.O. *Fallsch.Jäg.R.* 3	15/5/1944
96	*Oberst* Kroh, acting C.O. 2. Fallsch.Jäg.D.	12/9/1944
99	*General der Fallschirmtruppen* Ramcke, C.O. Fortress Brest	19/9/1944
112	*Oberst* Schulz, acting C.O. 1. Fallsch.Jäg.D.	18/11/1944
131	*Oberst* Walther, acting C.O. Fallsch.Pz.Gren.D. "H.G."	1/2/1945
155	*General der Fallschirmtruppen* Meindl, Com.Gen. II. Fallsch.K.	8/5/1945

In addition, 29 officers were awarded the Knight's Cross with Oak Leaves; 218 officers, non-commissioned officers and enlisted men were decorated with the Knight's Cross of the Iron Cross.

Divisions named in *Wehrmacht* communiqués 1943–45

1. Fallsch.Jäg.D.: 18/8, 24/12/1943; 25/3, 20/5, 29/6/1944
2. Fallsch.Jäg.D.: 10/9/1944
3. Fallsch.Jäg.D.: 19/6/1944
4. Fallsch.Jäg.D.: 2/6, 10/6, 28/6/1944.

(Does not include mention of division units and persons.)

3

Waffen-SS

History to the Outbreak of War

The history of this organization – it could not be considered a corps until 1933 – is inextricably linked with that of the National Socialist Party (NSDAP). When Hitler began reorganizing his party following his release from prison, he formed a group of young men, most of them veterans of the First World War, for his personal protection. It was dubbed the *Schutzstaffel* (protection unit), which was abbreviated to SS. The group's sole purpose was to provide personal protection for Hitler at election rallies, meetings, propaganda marches, etc.

In the years that followed, this modest protection unit was expanded into an independent organization which consciously sought to set itself apart from the *Sturmabteilungen* (SA). While its organization was similar to that of the SA, the men of the SS wore black uniforms.

In later years the former chief-of-staff of the *"Heeresgruppe Kronprinz"* (Army Group Crown Prince) in the First World War, *General* Count von der Schulenburg, wrote that the German Reich should create an elite corps which would be under the direct command of the Reich chancellor and fill the same role as the Kaiser's guards regiments of former years.

This concept was realized after Hitler assumed power in 1933. The formation of active SS units (*Bereitschaften*) with permanent quarters began the same year. In the beginning these units were most closely akin to police units.

The year 1933 also saw the formation of the *"Leibstandarte Adolf Hitler"*. In terms of personnel and armament, this formation was equivalent to a standard infantry regiment. Its members were young volunteers who had to meet strict height requirements and be in excellent health. The officer corps consisted almost entirely of men who had seen front-line action in the First World War.

The *"Leibstandarte"* (bodyguard), as it was referred to from then on, was quartered in the former Prussian cadet institute in Berlin-Lichterfelde. The *Standarte* received weapons and combat training at army troop training grounds in the Berlin area.

The men of the *"Leibstandarte"* wore the black uniform of the *"Allgemeine SS"* (general SS) with a band on the left sleeve bearing the title *"Adolf Hitler"*. Neither the armband nor the swastika band appeared on the uniform greatcoat.

The *Standarte*'s first leader – the designation commander was not used as yet – was *SS-Standartenführer* Dietrich, who had served in the First World War as a sergeant in the Royal Bavarian Tank Battalion.

The primary role of the *"Leibstandarte"* remained the personal protection of the Reich Chancellor, his ministers and visiting foreign dignitaries. In addition, the *Standarte* provided honor formations for national holidays and other similar events.

The year 1934 saw the establishment of the *SS-Junkerschule* (officer cadet school) in Bad Tölz and the formation of the *"Deutschland"* and *"Germania" Standarten* plus an *SS-Pioneer-Sturmbann* (equivalent to an army pioneer battalion). These units were based in Munich, Hamburg-Langenhorn and Dresden respectively.

On 16 March 1935 the *Reichstag* issued a law restoring German military sovereignty. Among its provisions was the expansion of the army to 36 divisions and the formation of a division of the armed SS. This law also provided for the establishment of a second officer cadet school in Brunswick and an *SS-Nachrichten-Sturmbann* (signals unit) in Unna, Westphalia.

The second *Junkerschule* was commanded by former *Generalleutnant* of the *Reichswehr* Hausser. In the years that followed, he became the trainer and educator of future *Waffen-SS* officers and the organizer of this new military force. To his soldiers he was "Papa Hausser".

The year 1935 brought the first public employment of the *"Leibstandarte"*. The unit, its men wearing their black uniforms with gray steel helmets, field packs and slung rifles, marched across the Saar bridges after the plebiscite held in that region.

The next two years saw the force aligned more closely with the army. Orders were issued in 1937 for the introduction of field-gray service uniforms for units of the armed SS and one year later for the wearing of shoulder straps and service colors with the field-gray uniform. In 1938 the *Reichsfuhrer-SS* established the *Inspektion der SS-Verfügungstruppe* (Inspectorate of Armed SS Units) and later ordered the formation of the *SS-Rgt. "Der Führer"* in the new defense districts in Ostmark (Austria).

In March 1938 the *Wehrmacht* had completed a bloodless occupation of Austria. It was an opportunity for the *"Leibstandarte"* to demonstrate its mobility and that it should be considered part of the army. At the beginning of March its was mobilized at its base in Berlin and ordered to Passau. There it joined *General* Guderian's XV. AK. (mot.). Together with the 13. PD., on 13 March 1938 it drove via Linz and St. Pölten to Vienna. The *"Leibstandarte"* covered 1 000 kilometers in two days. In October of the same year the *"Leibstandarte"* – its men now wearing field-gray army uniforms with the national eagle on the left sleeve and an "Adolf Hitler" cuff band – marched out of the Ore Mountains into the Sudetenland and occupied the area around Egerland.

On 17 August 1938 – before the entry into the Sudetenland – Hitler signed a decree which said:

> "...the *SS-Verfügungstruppe* is not a part of the armed forces or of the police, rather it is a standing armed force under my command..."

The decree also stated that the existing *SS Totenkopfverbände* (Death's Head Units), whose duty was to direct, administer and guard the concentration camps, was not an element of the *Wehrmacht* or the police, rather a standing unit of the SS.

The last year of peace saw the *SS-Verfügungstruppe* completely motorized and equipped with heavy weapons. As well, an artillery regiment was formed in Jüterbog.

The *SS-Verfügungstruppe*'s peacetime organization in the late summer of 1939 was as follows:

SS-Leibstandarte with three motorized infantry battalions, 13th to 15th Companies, pioneer platoon, armored scout platoon, guard battalion in Berlin.
SS-Standarte "Deutschland" (organized as above without pioneer platoon) in Munich.
SS-Standarte "Germania" (organized as above) in Hamburg-Langenhorn, Arolsen, Ellwangen.
SS-Standarte "Der Führer" (organized as above) in Vienna, Graz, Klagenfurt.
SS-Standarte z.b.V. with two motorcycle battalions, signals platoon in Ellwangen and Dachau.
SS-AR (mot.) with three battalion of light field howitzers in Jüterbog.
SS-Pi.Sturmb. with three companies, one light transport column in Dresden.
SS-Nachr.Sturmb. with two companies, one light transport column in Unna.

The Second World War

On 26 August 1939 the "Secret Mobilization" order was issued, sending the SS units into the Reich's eastern regions. The *SS-Leibstandarte*, the *Regiment "Deutschland"*, the SS-Artillery Regiment, the SS-Signals *Sturmbann* and the newly-formed SS supply units and SS medical company had already been sent to East Prussia on the pretext of taking part in the 25th anniversary of the Battle of Tannenberg. The SS Special Duties *Standarte* in Ellwangen was disbanded and the two motorcycle battalions formed the new *SS-Aufkl.Abt.* (reconnaissance battalion), which was also dispatched to East Prussia.

These units were attached to the newly-formed *Pz.Verband Ostpreußen* (Armored Group East Prussia) commanded by *Generalmajor* Kempf. SS-Gruf. Hausser, who also went to East Prussia, was given no command role, instead he remained with the troops as an observer. The named units took part in the armored group's actions on the right wing of the German *3. Armee* in East Prussia until Poland's surrender.

When war began, the remaining SS regiments, battalions and independent battalions came under the command of *Armee* headquarters: *"Germania"* to the AOK 14 in southern Poland, *"Der Führer"* to the AOK 7 in Bavaria and the *SS-Pi.Sturmb.* and the *"Leibstandarte"* to the AOK 10 in Lower Silesia.

When war broke out, the *SS-Totenkopf* units in East Prussia were concentrated in the Danzig area to form the *SS-Heimwehr "Danzig"* (SS Home Defense Danzig), a formation roughly equivalent to a regiment. This force took part in the fighting in and around Danzig. In those days it was under the command of *Generalmajor* Eberhardt's brigade made up of army units.

After the Polish campaign the SS units were reorganized and new units formed. After its return to Berlin, the *"Leibstandarte"* was re-equipped as a reinforced motorized infantry regiment. The three formerly independent regiments *"Germania"*, *"Der Führer"* and *"Deutschland"* were combined with newly-created division units to form the *SS-Verfügungsdivision* (SS-Verf.D.). Formation of the *SS-Totenkopfdivision* and the *SS-Polizeidivision* began in October 1939. As well, the weeks that followed saw the formation of ten independent SS infantry regiments, an SS cavalry regiment (Warsaw), an SS signals battalion and SS infantry and non-commissioned officers schools in Radolfzell and Lauenburg, Pomerania.

The officers, non-commissioned officers and enlisted men for these new units came in part from the *SS-Totenkopfstandarten* (concentration camp guard units) as well as from police organizations plus numerous volunteers. At the start of the western campaign *SS-Ersatz-Sturmbanne* (SS replacement units) were formed in Berlin-Lichterfelde, Hamburg-Langenhorn, Breslau, Arolsen and Brunswick. As well, the *Pi.-Ersatz-Btl.* (pioneer replacement battalion) was formed in Dresden along with the *Nachr.Ersatz-Btl.* (signals replacement battalion). The *SS-Junkerschule* formerly located in Brunswick was moved to Vienna.

When the western campaign began on 10 May 1940 (no SS units took part in the fighting in Norway), the *SS-Verfügungsdivision* and the *"Leibstandarte"* were in the front lines. The *SS-Verfügungsdivision* saw action in the Netherlands in the initial phase of the campaign, while the *"Leibstandarte"* operated as part of *Panzergruppe Guderian*. Ten days after the campaign began, the *SS-Totenkopfdivision* arrived. The *SS-*

Polizeidivision manned positions in the *Westwall* on the Upper Rhine. Both SS divisions and the *"Leibstandarte"* took part in the fighting in the Netherlands, Belgium and northern France (Battle of Dunkirk). The second phase of the campaign brought the *SS-Verfügungsdivision* and the *SS-Totenkopfdivision* to the Spanish border, the *"Leibstandarte"* to Lyon and the *SS-Polizeidivision* to southwestern France.

After the western campaign the *SS-Division "Wiking"* was formed from numerous volunteers from the Reich, Denmark, Norway, Finland, Sweden and the Netherlands. Volunteers also came from France and especially from Spain. The latter served mainly in units of the army ("Blue Division"). Not until 1942 were all volunteers from southeastern and eastern Europe concentrated exclusively in units of the *Waffen-SS*.

The term *Waffen-SS*, which henceforth was the designation for the armed SS units, was first coined by Hitler himself. In his noteworthy speech to the *Reichstag* in Berlin on 19 July 1940 he said:

> "My deputies, in this struggle the German foot soldiers have again proved to be what they always were: the best infantry in the world. Emulating them are all other branches of the army, especially the young units of the armored and motorized troops. The men of the *Waffen-SS* share this glory!"

Four officers and one non-commissioned officer were decorated with the Knight's Cross after the 1940 campaign.

There now followed a rapid expansion of the units. The *"Leibstandarte"* had been expanded into a motorized brigade soon after the western campaign. Spring 1941 saw the formation of the *SS-Division "Wiking"*, a motorized division made up of volunteers from northern and western Europe.

Only the *SS-Polizeidivision*, made up of members of the police and army units of *Wehrkreis III* (Berlin), did not come under the command of the SS command offices. Instead, until the start of the eastern campaign, it was subordinate to the Chief of the Order Police (*Ordnungspolizei*) in the Reich Ministry of the Interior. Until 1943 the units of this division wore the field gray uniform of the police, but with SS rank badges. The previous police rank structure was retained.

Before these changes were completed, however, a new theater of war came into being: the Balkans. The *SS-Verfügungsdivision* and the

Brig. Leibstandarte took part in this campaign. The first division was attached to the *XXXXI. AK.mot.* of *Pz.Gruppe Kleist*, which advanced into Yugoslavia from Romania with the objective of Belgrade. An advance detachment of the *2./Kradschtz.Btl. "Das Reich"* under SS-Hpt.Sturmf. Klingenberg was the first element of the armored group to reach Belgrade on 12 April.

The *"Leibstandarte"* was part of the XI. AK. of the *12. Armee*. This corps formed the western group of the army, which advanced into Greece from the north. The *"Leibstandarte"* advanced through the Klisura Pass toward Grevena. There the entire Greek army was surrounded. After its surrender SS-Ob.Gruf. Dietrich led the surrender negotiations on the German side.

The German buildup against Russia proceeded even while the fighting raged in the Balkans. The first units of the *Waffen-SS* and the *SS-Polizeidivision* arrived on the Reich's eastern border by mid-April 1941. When the campaign against the Red Army began on 22 June 1941, the following units were in the front or in reserve close behind:

> *"Leibstandarte"* and *SS-Division "Wiking"* with the *6. Armee*;
> *SS-Division "Das Reich"* with *Pz.Gruppe 2* (XXXXVI. AK.);
> *SS-Division "Totenkopf"* with *Pz.Gruppe 3* (XXXXI. AK.);
> *SS-Polizeidivision* OKH reserve (L. AK.).

The three motorized divisions and the *"Leibstandarte"* took part in the war against the Soviet Union from the very first day of the campaign. Units of the *Waffen-SS* were in the forefront of the advance by their respective army groups. Thanks to their excellent equipment and vehicles they were able to keep pace with the army's motorized units and often surpass them.

A summary of their operations:

> Army Group South: in the beginning the two *Waffen-SS* units were held in reserve, but were soon committed in the front lines. My mid-July the *"Leibstandarte"* was far ahead of the other divisions of the army group and was closest to the Ukrainian capital of Kiev; however, it was turned southwest to close the pocket at Uman. One month later the *SS-Division "Wiking"* took Dnepropetrovsk and the

"Leibstandarte" took Khersonyes at the mouth of the Dniepr. The *SS-Division "Wiking"* and the *SS-Division "Das Reich"*, which had arrived in the central sector, took part in the battle of encirclement at Kiev, while the *"Leibstandarte"* – now enlarged to the size of a panzer division – advanced to the Sea of Azov, taking Taganrog and Rostov at the mouth of the Danube.

Army Group Center: The *SS-Division "Das Reich"* was part of *Panzergruppe Guderian* and with it participated in the battle of encirclement at Minsk, later advancing into the Smolensk area. Around Yelnya it fought a fierce defensive battle against far-superior forces of the Red Army. When the Panzergruppe turned south to take part in the battle of encirclement at Kiev, the division crossed the Desna near Paritschi and advanced to east of Priluki. A few days before the end of the battle the division was moved northeast to take part in "Operation Typhoon", the attack on Moscow. The division was committed at the *4. Pz.Armee*'s point of main effort and was one of the divisions closest to Moscow. In the meantime, the newly-created *SS-Kav.Brig. 1* had arrived in the rear areas of Army Groups South and Center and saw action in the anti-partisan role in the Pripyat Marshes.

Army Group North: the *SS-Division "Totenkopf"* was attached to the *16. Armee* and in the first days of the campaign it advanced across the Dvina northeast in the direction of Lake Ilmen. By the end of the year the units had reached the wooded, marshy area of the Waldai Mountains southeast of the lake. The sudden onset of winter prevented a further advance. The *SS-Polizeidivision*, part of the *18. Armee* and for a time of the *4. Pz.Armee*, took part in the advance through the Baltic states and northern Russia to the southern suburbs of Leningrad. The division, at times committed in battle groups, suffered heavy losses.

The *Wehrmacht*'s successes against the Soviet Union in the first year the war motivated many Germans and non-Germans to join the ranks of the *Waffen-SS*. As a result, the number of soldiers in this corps rose from 220,000 to 330,000 in the years 1942 and 1943. In 1941 the foreign volunteers were concentrated in so-called "SS Legions". In 1942 these included the *SS-Legionen "Flandern"*, *"Wallonien"*, *"Nederlande"* and *"Frankreich"*. As well there was the *Freikorps Danmark"* and the two

SS battalions *"Estland"* and *"Lettland"*. In 1942 these new units saw action exclusively on the Eastern Front, either as part of existing divisions or independently.

In the east in 1942 the divisions, brigades, legions and independent formations of the *Waffen-SS* were in the heart of the action. In the north the *SS-Polizeidivision* continued to fight at Leningrad and on the Volkhov, the *SS-Division "Totenkopf"* in the Demyansk pocket, and the Baltic legions before Leningrad and on the Volkhov. The *SS-Division "Das Reich"* remained in the central sector and in the southern sector the *SS-Division "Wiking"* crossed the Don near Rostov and by the end of August fought its way toward the mountains of the Caucasus. The *"Leibstandarte"* had been transferred to France in July for reorganization into a panzer division.

At the beginning of 1943 the *Waffen-SS* found itself in the same situation as the army, navy and air force: it was no longer possible to expand its forces by relying solely on volunteers. SS recruiters sometimes exerted pressure on those eligible for military service and forced ethnic Germans in the occupied territories to join the *Waffen-SS*. Recruitment was expanded to include non-German peoples such as Albanians, Estonians, Finns, Galicians, Cossacks, Latvians, Lithuanians, Moslems, Russians, Ukrainians and White Ruthenes.

The composition of the new *Waffen-SS* divisions created in 1942-43 is reflected in the designations given major units. The composition of individual divisions was expressed in their designations, especially after 1943, when the major expansion of the *Waffen-SS* began. Divisions which consisted mainly of volunteers from the "Germanic" race were designated *"Freiwillige D"*. Divisions with soldiers from other ethnic groups were designated *"Waffen-Gren D.-SS"*.

After 1942 the formation of new *Waffen-SS* divisions followed roughly the same pattern as those of the army, however the *Waffen-SS* units were often more powerfully armed and organized than their army counterparts. This state of affairs could be justified to some extent by the fact that the *Waffen-SS* divisions were often employed as "fire-brigades" in extremely difficult military situations.

In the course of the expansion of the *Waffen-SS*, in July 1942 the first corps headquarters was formed with the designation SS-Pz.K. The headquarters was put together with the corresponding corps units at the Bergen troop training grounds. The corps' first commanding general was *SS-Ob.Gruf.* Hausser. In December 1942 the corps headquarters transferred to Toulon in southern France, however in January 1943 it was summoned to the Eastern Front.

The corps headquarters assumed command in a sector of front in the Kharkov area but, faced with the imminent encirclement of the city, on 15 February 1943 it evacuated Kharkov. The Red Army was advancing westward. Contrary to Hitler's orders, the corps' three divisions (*"Leibstandarte"*, *"Das Reich"* and *"Totenkopf"*) did not retire but instead went round Kharkov to the east and attacked the city, which was recaptured on 14 March 1943. This was the first German success in the east in a long time.

The second combined employment of the three divisions was in "Operation Zitadelle", the last major German offensive in the east. The objective of the operation was to pinch off the Soviet salient near Kursk, which projected like a boil into the German front. Army Group South, in combination with Army Group North attacking from the north, was supposed to eliminate this salient.

The attack by the two German army groups began on 5 July 1943. In the south the II. SS-Pz.K. (as the SS.Pz.K. had been renamed in June) with its three panzer divisions formed the right wing of the *4. Pz.Armee*. On the second day of the offensive the three divisions were able to breach the enemy's main line of resistance and advance up to 18 kilometers to the northeast. However, the Red Army, which had known about Zitadelle for a long time, had readied its tank forces for a counterattack. In mid-July the Soviets counterattacked, resulting in the biggest tank battle of the Second World War near Prokhorovka. The three SS panzer divisions formed the German side. Then the corps and the entire army group had to break off the attack.

In the period from 5 to 19 July 1943 the II. SS-Pz.K. lost a total of 1,447 killed, 6,198 wounded and 138 missing (prisoners). The three divisions of the corps, the first of the *Waffen-SS*, were subsequently split up and fought in various sectors, sometimes under the command of army corps headquarters.

By the end of 1943 the *Waffen-SS* numbered seven corps headquarters:

The I. SS-Pz.K. had been formed at Berlin-Lichterfelde on 27 July 1943, the II. SS-Pz.K. already existed, the III. SS-Pz.K. was created in Grafenwöhr on 15 April 1943, the IV. SS-Pz.K. was formed in Poitiers on 30 August 1943 (this formation was aborted and the corps headquarters became the VI. SS-K.), the V. SS-Geb.K. was formed in Berlin, the VI. SS-K. came together at Grafenwöhr on 8 October 1943, and the VII. SS-Pz.K. was established on 3 October 1943.

1943 also saw the creation of many new divisions and brigades. The number of officers, non-commissioned officers and enlisted men rose to 540,000 by the end of 1943 and to 910,000 one year later. As a result, the *Waffen-SS* could no longer be considered an "elite corps".

The latter figure includes 40,000 former members of the *Luftwaffe* from disbanded air units or *Luftwaffe* field divisions plus 5,000 former members of the *Kriegsmarine* from disbanded units.

On 22 October 1943 the *Reichsführer-SS* ordered the numbering of all units of the *Waffen-SS* down to the independent *Sturmbanne* (battalion strength). In action at the end of 1943 were:

1. SS-PD. *"Leibstandarte"*	Eastern Front
2. SS-PD. *"Das Reich"*	Eastern Front
3. SS-PD. *"Totenkopf"*	Eastern Front
4. SS-Pol.D.	Eastern Front
5. SS-PD. *"Wiking"*	Eastern Front
6. SS-Geb.D. *"Nord"*	Finland
7. SS-Geb.D. *"Prinz Eugen"*	Balkans
8. SS-Kav.D. *"Florian Geyer"*	Eastern Front
9. SS-PD. *"Hohenstaufen"*	France
10. SS-PD. *"Frundsberg"*	France
11. SS-Pz.Gren.D. *"Nordland"*	Eastern Front
12. SS-PD. *"Hitlerjugend"*	France
13. SS-Geb.D. *"Handschar"*	Home Front
14. SS-Gren.D. (*Galizien 1*)	Home Front
15. SS-Gren.D. (*Lettland 1*)	Eastern Front
16. SS-Pz.Gren.D. *"Reichsführer-SS"*	France
17. SS-Pz.Gren.D. *"Götz von Berlichingen"*	France

Thus the bulk of the seventeen large formations created to date (not counting independent brigades, regiments and special units) were in action on the Eastern Front. There were also six division headquarters with the designation *"Befehlshaber der Waffen-SS"* (commander of the *Waffen-SS*). These command agencies were responsible for the technical services, schools and replacement troops of the *Waffen-SS*. Their jurisdiction was limited to the rear areas in territories occupied by the *Wehrmacht*. These were Bohemia-Moravia, Finland, *General-Gouvernement* (Poland), Holland, Norway and Hungary.

The year 1944 saw units of the *Waffen-SS* committed on every front in Europe. Not until the middle of the year, after the Allied invasion of France, did the best divisions in that area engage British, Canadian and American army troops. Losses to enemy tactical aircraft were heavy.

On the day of the invasion (6 June 1944) the 12. SS-Panzer-Division received orders to proceed to Rouen. Exactly twenty-four hours later the 17. SS-Pz.Gren.D. was ordered from the Avranches area to St. Lô and Carentan, the focal point of the American attack. That day the tanks of the 12. SS-PD. destroyed 28 enemy tanks. The 2., 9. and 10. SS-PD. reached the battlefield by the end of June, the 1. SS-PD. somewhat later.

The bulk of these units, attached to the 7. *Armee* under SS-Ob.Gruf. Hausser, were trapped in the Falaise Pocket at the beginning of August.

The 12. SS-PD., which included some of the youngest soldiers in the *Wehrmacht*, distinguished itself in combat. On 8 August it halted the first attack by the Canadian II Corps, destroying 47 enemy tanks. The German side also suffered heavy losses. The 12. SS-PD. destroyed 278 enemy armored vehicles in a few days, however it was itself reduced to no more than 500 men with about 20 tanks and few guns.

A British general later wrote:

> "The SS division 'Hitlerjugend' fought with a determination and fury which was never encountered again during the entire campaign."

When the *7. Armee* was surrounded at Falaise on 15 August 1944, inside the pocket was the I. SS-Pz.K. and from the right wing of the surrounded *Armee* in the area north of Falaise to the left wing between Argentan and Chambois the 12. SS-PD., 9. SS-PD., 10. SS-PD., 2. SS-PD. and 17. SS-Pz.Gren.D. held the front between army and parachute divisions. Two days later, when the *7. Armee* broke out to the east, the last 60 men of the 12. SS-PD. held out in Falaise until none were left alive.

The road to the frontiers of the Reich was open.

Acting on direction from the *Reichsführer-SS*, the command office of the *Waffen-SS*, the *SS-Führungshauptamt*, formed new commands. In September the AOK of the *6. SS-Pz.Armee* was created, plus corps headquarters of the IX., XI., XII., XIII. and XIV. SS.K. (the IX. was designated a mountain corps).

After the collapse of the Eastern Front, the Soviet forces were not halted until the East Prussian border, the Warsaw area and before Upper Silesia. The collapse of the Western Front drove German troops back towards the Reich border with the exception of the Netherlands and Alsace-Lorraine.

The next battlefield was to be the area between Eindhoven and Arnhem, between the Maas, Waal and Rhine Rivers. There, on 16 September 1944, the Allies launched "Operation Market Garden", an airborne landing behind the German front by two American, one British and one Polish airborne divisions. The German counterattack, which included the XII. SS.K. with the 9. and 10. SS-PD., smashed the Allied airborne operation in less than two weeks.

The following list depicts where the divisions of the *Waffen-SS* were in action on 9 November 1944:

1. SS-PD.	reformation in the Siegburg area
2. SS-PD.	readiness position in the Eifel
3. SS-PD.	in action at Warsaw
4. SS-Pz.Gren.D.	in action near Szolnok, Hungary
5. SS-PD.	in action at Modlin and Warsaw
6. SS-Geb.D.	transport from Oslo to the Saarpfalz
7. SS-Geb.D.	in action at Nisch-Kraljevo
8. SS-Kav.D.	in action in and around Budapest
9. SS-PD.	rest and refit in the Siegen area
10. SS-PD.	in action between Jülich and Aachen
11. SS-Pz.Gren.D.	in action in Courland
12. SS-PD.	rest and refit in Weserland
14. SS-Gren.D.	new formation in Neuhammer
15. SS-Gren.D.	new formation around Konitz

16. SS-Pz.Gren.D.	in action in northern Italy
17. SS-Pz.Gren.D.	in action in Saar region
18. SS-Pz.Gren.D.	formation in the Batschka
19. SS-Gren.D.	in action in Courland
20. SS-Gren.D.	new formation in Neuhammer
21. SS-Geb.D.	in action at the Drina
22. SS-Kav.D.	in action in and around Budapest
25. SS-Gren.D.	formation in western Hungary
27. SS-Gren.D.	formation in the Lüneburg Heath
28. SS-Pz.Gren.D.	rest and refit in southern Hanover
29. SS-Gren.D.	in action in northern Italy
30. SS-Gren.D.	in action on the Upper Rhine front
31. SS-Gren.D.	in action at Lake Balaton
34. SS-Gren.Br.	formation in the Netherlands
36. SS-Pol.Br.	anti-partisan action in Slovakia
Indische Legion	formation Königsbrück
Brigade Dirlewanger	in action in Warsaw
Osttürkische Standarte	formation in Slovakia

The Ardennes offensive, which the German command hoped would bring about a turn in Germany's fortunes, was the last major action involving units of the *Waffen-SS*. The newly-created *6. SS-Pz.Armee* under the command of SS-Ob.Gruf. Dietrich was to play a major role in the offensive. Operating on the right wing of *Feldmarschall* von Rundstedt's army group, it was supposed to shatter the Allied front in the High Eifel and drive through Malmédy to Antwerp.

On its right wing the *Armee* committed the LXVII. AK. with four infantry divisions and one parachute division. The I. SS-Pz.K. (12. and 1. SS-PD. and the 150. SS-Pz.Brig.) was to break through the enemy front and drive towards Malmédy. The II. SS-Pz.K. (2. and 9. SS-PD.) would then drive through the breach and advance to the northwest.

Contrary to expectations, the snow- and ice-covered high country of the Eifel represented an almost impossible obstacle to the heavy tanks. Enemy resistance was also fiercer than expected. The I. SS-Pz.K. became bogged down between Elsenborn and Malmédy. The II. SS-Pz.K. was able to gain ground through the forests to the Ourthe and the 1. SS-PD. occupied Stavelot, however by that point the units were exhausted.

When the weather improved the Allied air forces struck. During the Christmas season 1944 the German offensive strength petered out and *Feldmarschall* von Rundstedt was forced to order a retreat behind the border.

Hitler was disappointed by the failure of his SS units. He subsequently ordered the *6. SS-Pz.Armee* to Hungary. There, in January 1945, it received orders to lead the advance toward Budapest. Powerful forces of the Red Army had surrounded the Hungarian capital. The city was defended by one panzer, one panzer-grenadier and one infantry division of the army plus the 8. and 22. SS-Kav.D.

The *6. SS-Pz.Armee*'s attack from the northwest failed to get through; then the *Armee* – which included the 3. and 5. SS-PD – attacked from the southwest. This attack brought the two SS panzer divisions and the army's 3. PD to the Danube but no farther. Hitler was so enraged that he ordered the *"Leibstandarte"* to remove its cuff titles which bore his name.

The *6. SS-Pz.Armee* subsequently joined the general retreat of the German forces in the east. The I. and II. SS-Pz.K. under its command (1., 2., 3. and 12. SS-PD.) withdrew into the Vienna area where they surrendered.

In the final weeks of the war the remaining divisions of the *Waffen-SS* were in every sector of the front, from Pomerania to Saxony, from the North Sea to the Alps, in Italy and the Balkans. The 11. SS-Pz.Gren.D. was one of the units which allowed itself to be encircled in Berlin. Its ranks included volunteers from France, Norway, the Netherlands and Spain and they fought on until the "Red Banner" fluttered over the shattered city.

The *Waffen-SS* suffered a rate of loss second only to the German submarine arm – that should never be forgotten!

Senior Commands

SS-Führungshauptamt

The *SS-Führungshauptamt* was the supreme command authority of the *Waffen-SS* in the home war zone, roughly comparable to the office of the Commander of the Replacement Army.
Command list on 9 November 1944:

Chief of the *SS-Führungshauptamt*: O.Gruf. Jüttner
Chief of the Command Office: Gruf. Petri
(in charge of: Administration Office, Office for Cavalry and Horse-Drawn Transport, Office for Motor Transport, Supply Office, Office for Officer and Non-Commissioned Officer Training with 4 officer candidate, 2 NCO and 9 service schools plus 2 troop training grounds)
Chief of Inspectorate of Training: O.Gruf. Knoblauch
Chief of Inspectorate of Signals: Brig.F. Keilhauer
Chief of Inspectorate of Medical Services: Gruf. Dr. Genzken.

High Command of the *6. SS-Panzerarmee*

The high command of this single *Armee* of the *Waffen-SS* was created on 24 September 1944 in *Wehrkreis VI* from the headquarters of the former *Wehrmacht* Commander Belgium-Northern France, the Gen.Kdo. XII. AK and elements of the *Waffen-SS*.

Command list on 10 December 1944:
Commander-in-Chief: Ob.Gruf. Dietrich
Chief-of-Staff: Brig.F. Kraemer
Ia (Operations Officer): Ob.Sturmb.F. Mayer

Corps Headquarters of the *Waffen-SS*

Date of formation and commanding general on 10 December 1944:

I. SS-Pz.K.: Formation 27 July 1943 in Berlin-Lichterfelde,
Commanding general: O.Gruf. Keppler;
II. SS-Pz.K.: Formation in June 1943 in Bergen as SS-Pz.K., renamed II. SS-Pz.K. in June 1943.
Commanding general: O.Gruf. Bittrich;
III. SS-Pz.K.: Formation 15 April 1943 in Grafenwöhr,
Commanding general: O.Gruf. Steiner;
IV. SS-Pz.K.: Formation 30 August 1943 in Poitiers,
Commanding general: O.Gruf. Gille;
V. SS-Geb.K.: Formation on 1 July 1943 in Berlin,
Commanding general: O.Gruf. Krüger, F.W.;
VI. SS-AK.: Formation on 8 October 1943 in Grafenwöhr,
Commanding general: O.Gruf. Krüger, W.;
VII. SS-Pz.K.: Formation on 3 October 1943 in Berlin, renamed IV. SS-Pz.K. (new) on 30 June 1944,

IX. SS-Geb.K.: Formation in July 1944 in Croatia,
Commanding general: O.Gruf. Sauberzweig;
X. SS-AK.: Formation in January 1945 from XIV. SS-AK.,
disbanded without seeing action;
XI. SS-AK.: Formation on 24 July 1944 in Ottomachau,
Commanding general: O.Gruf. Kleinheisterkamp;
XII. SS-AK.: Formation on 1 August 1944 in Silesia,
Commanding general: O.Gruf. von Gottberg;
XIII. SS-AK.: Formation on 1 August 1944 in Breslau,
Commanding general: Gruf. Simon;
XIV. SS-AK.: Formation in November 1944 on the Upper Rhine, renamed X. SS-AK. in January 1945;
XV. SS-Kosaken-K. Formation on 1 February 1945 in Croatia,
Commanding general: Glt. von Pannwitz;
XVIII. SS-AK.: Formation in December 1944 on the Upper Rhine,
Commanding general: Gruf. Reinefarth.

The highest-ranking officer of the *Waffen-SS* at that time was Ob.Gruf. Hausser, Commander-in-Chief of Army Group G in Alsace-Lorraine.

All ranks were preceded by the prefix SS- (for example, SS-Ob.Gruf.); this prefix has been omitted in this work.

The Divisions

1. SS-Panzerdivision "Adolf Hitler"

The division was the successor of the first armed unit of what would become the *Waffen-SS*. The unit was created after Hitler's assumption of power with the title *"Leibstandarte Adolf Hitler"*. Military training of the regiment-size unit took place in the barracks installations of the former Prussian Cadet Institute in Berlin-Lichterfelde and at the army troop training grounds in *Wehrkreis III* (Berlin).

When the Second World War began, the fully-motorized *Standarte* was attached to the high command of the *10. Armee* as operational reserve. In the Polish campaign it marched out of the Oels area behind the 17. ID. and later the 4. PD. across the Warthe and Bzura Rivers to the area around Modlin and Warsaw.

The western campaign saw the *"Leibstandarte"* depart the Rheine area as part of the 227. ID. and advance through Holland to the Scheveningen area. Subsequently attached to the XXXIX. AK.mot. of *General* Guderian's panzer group, the unit took part in the attack through Arras toward Boulogne and Dunkirk, distinguishing itself several times in the process. The second phase of the 1940 western campaign took the *"Leibstandarte"* across the Somme, the Marne and the Seine to the area west of Lyon, still attached to *Panzergruppe Guderian*.

After the western campaign, in August 1940 the *Standarte* was enlarged into a motorized brigade, retaining its *"Leibstandarte"* (bodyguard) designation.

The brigade's first major action began on 6 April 1941 in the Balkan campaign. The brigade set out from the area around Küstendil in Bulgaria, attacking through Skolpje, Prilep, Bitolj, Petras and Pyrgos to Tripoli. There the commander of the *"Leibstandarte"* took part in the negotiations for the surrender of the Greek army.

A few days later the brigade was transferred to southern Poland. From there, at the beginning of July 1941, it began the advance into southern Russia under the command of *Panzergruppe von Kleist* (later the *1. Pz.Armee*). Often employed as the *Panzergruppe*'s spearhead unit, the *"Leibstandarte"* advanced through Luczk and Dubno in the direction of Zwiahel and Zhitomir and was the first large unit of the *Wehrmacht* to reach the area west of Kiev. There the *"Leibstandarte"* was forced to make an about face to support the infantry units in the battle of encirclement at Uman. From August 1941 the rapid advance continued through Konstantinovka and Pervomaysk to the Dniepr, which was crossed near Berislav. After brief defensive engagements near Kamenka and Balki, the brigade crossed the Mius south of Stalino as part of the XIV. Pz.K. By the beginning of December the motorized columns had advanced to north of Rostov, but then the start of the Russian winter offensive resulted in a withdrawal to the Mius. The *"Leibstandarte"* remained there until the end of May 1942, living and fighting in trenches.

The brigade was subsequently transferred to Normandy where, on 15 July 1942, it was converted into the SS-Pz.Gren.D. *"Leibstandarte"*. The conversion kept the new division occupied until January 1943. It was organized as follows:

SS-Pz.Gren.R. 1 and 2, plus panzer and artillery regiments, reconnaissance battalion, anti-tank battalion, assault gun battalion, flak and signals battalions, a pioneer battalion and supply units.

In February 1943 the new SS-Pz.Gren.D. *"Leibstandarte"* was transferred to Army Group B on the southern front. There it took part in the Battle of Kharkov and the daring attack which recaptured the Ukrainian capital. After the battle the division remained in the Kharkov area until the beginning of June. Then began "Operation Zitadelle". As the shock force of the 4. Pz.Armee, the SS-Pz.Gren.D. *"Leibstandarte"* broke through the Russian trench lines in heavy fighting and advanced to Prokhorovka, where the greatest tank battle of the Second World War was fought. The battalions of the *"Leibstandarte"* fought the massed tank forces of the Soviet II and V Guards Tank Armies. But the Soviets were stronger. In mid-July Russian armies broke through to Orel and attempted to encircle Army Group Center; as a result the Battle of Kursk was broken off.

The *"Leibstandarte"* was immediately pulled out of the front and rushed to northern Italy to disarm the Italian Army after the fall of Mussolini and Marshal Badoglio's peace offer to the Allies.

The entire II. SS-Pz.K. arrived in northern Italy. The units of the *"Leibstandarte"* assembled in Tyrol then crossed the Brenner to Milan and northern Italy. There, in the period to mid-September, 106,000 Italian troops were disarmed in the area of Modena, Bologna, Florence, Padua, Verona and Milan.

After the restoration of general order in northern Italy, the division was able to turn its attention to a reorganization. At the end of September the reinforced 1. SS-Pz.Gren.R. was temporarily employed against Serbian partisans.

Effective 22 October 1943 (all divisions, regiments and battalions of the *Waffen-SS* were numbered on this day), the *"Leibstandarte"* was renamed the 1. SS-PD. The units attached to the division also received the number 1 (the rocket battalion and trained replacement battalion assigned to the division in September and October 1944 also received the number 1).

At the end of October the division was again loaded aboard trains, and by November it arrived in the area of Zhitomir. There it was attached to the *4. Pz.Armee*. Until the end of year the men of the *"Leibstandarte"* took part in fighting at Brusilov, north of Radomyschl, at Chepovichi and west of Berdichev.

The new year of 1944 saw the division still in the southern sector of the Eastern Front. The *"Leibstandarte"* took part in the fighting in the Zvenigorodka, Pogrebishchev and Vinnitsa areas. Then the division was committed to the effort to relieve the divisions surrounded at Cherkassy. The division was itself encircled with the *1. Pz.Armee* in the area around Kamenets-Podolsk (Hube Pocket). With the *1. Pz.Armee* it broke out through Vinnitsa and Proskurov to Ternopol.

The *"Leibstandarte"* halted there, after which it was transferred to Belgium to rest and reequip. There was little time, however, for on 6 June 1944 the Allied invasion of Normandy began. The division's next deployment was predetermined.

The *"Leibstandarte"* arrived at the front around Caen in the first days of July. There, together with the 12. SS-PD. and panzer and infantry divisions of the army, it halted an attack by the British 2nd Army.

The Allies subsequently broke through at Avranches and trapped the German *7. Armee* at Falaise. The *"Leibstandarte"* was among the units in the pocket. The division was ordered to spearhead a breakout in the direction of Chambois. The subsequent retreat took the division to the Aachen area.

The *"Leibstandarte"* was withdrawn from the front and transferred to Westphalia to reequip. The division's regiments and independent battalions were brought up to full strength and the panzer regiment was issued the latest *Königstiger* (King Tiger) tanks. After a brief training period the division returned to the front in the Eifel.

The newly-formed *6. SS-Pz.Armee* was to play a vital role in the coming Ardennes offensive. Infantry divisions were to crack the Allied front between Monschau and Elsenborn, then the four panzer divisions of the I. and II. SS-Pz.K. were to drive through the gap. The 1. SS-PD. received orders to advance via Malmédy and the Maas to Antwerp. Parachute troops were supposed to open the roads through the forests for the armored forces. The infantry units made little progress on the first day and the panzer-grenadiers of the 1. SS-PD had to open the way for the panzer regiment. The dense forests and poor roads hindered the advance and the King Tigers saw little action.

Finally, after several days of costly fighting, the companies of SS-PR. 1 reached Stavelot and La Gleize. There the advance was stopped and on Christmas Day 1944 the battle had to be broken off. The last German offensive of the Second World War had failed.

The entire *6. SS-Pz.Armee* was pulled out of the front and at the end of January 1945 it was rushed to Hungary to relieve Budapest, which was under siege by the Red Army. This attack, too, failed. From March 1945 the 1. SS-PD. was engaged in fighting at Lake Balaton and near Simontornys. Then it retreated through Veszprem, Varoslöd and Herend into the area west of Vienna. There the battle groups of the *"Leibstandarte"* surrendered on 8 May 1945.

Commanders:
O.Gruf. Dietrich; Brig.F. Wisch; Brig.F. Kumm; Ob.F. Mohnke.

2. SS-Panzerdivision "Das Reich"

In August 1939 the units of the *SS-Verfügungstruppe* which were scattered among various garrisons in the German Reich were placed on alert and transferred to East Prussia. When the Polish campaign began on 1 September 1939 these units were attached to the *Pz.Verband Ostpreußen* (Armored Unit East Prussia) under *Generalmajor* Kempf. The unit assembled for the attack north of Mlawa. A frontal attack against the fortress of Mlawa failed with heavy losses. The unit was then moved south via Rozan and reached the area southwest of Siedlce.

After the campaign the four *SS-Standarten "Deutschland"*, *"Das Reich"*, *"Germania"* and *"Der Führer"* were brought together with the *Art.Standarte* and the signals, pioneer and anti-tank *Sturmbanne* plus supply units in western Germany. The so called *SS Verfügungsdivision*, the first large unit of the *Waffen-SS*, was created there on 1 April 1940. In command of the division was Gruf. Hausser.

When the western campaign began on 10 May 1940 the new division spearheaded the attack by the *18. Armee* and on the first day of the campaign captured the Dutch forts of the Grebbe Line and occupied Arnhem. On the third day the *"Germania"* Regiment reached the Dutch-Belgian border near Antwerp and fought its way against French troops across the Zuid-Beverland isthmus to Vlissingen. In the last days of May the division took part in the Battle of Dunkirk after advancing between Lille and Cassel.

In the second phase of the campaign, on 14 June the division crossed the Marne. From there it advanced deep into southwestern France and on 21 June 1940 reached the Spanish border. There the *Standarte "Das Reich"* had to be released for the formation of other large units. In return the division was fully motorized.

In autumn 1940 the *SS-Verfügungsdivision* was renamed the *SS-Division "Das Reich"*. When the campaign in the Balkans began on 6 April 1941 the division was attached to Pz.Gruppe von Kleist. The units of the division spearheaded the *Panzergruppe* and were the first of the *Wehrmacht* to enter Sarajevo and Belgrade.

During the final weeks of the campaign the division was put aboard trains and transferred to the Reich's eastern border. It was attached to *Pz.Gruppe Guderian*, later to the *2. Pz.Armee*. As part of the XXXXVI. AK.mot., the division crossed the Bug north of Wlodawa and after three weeks of fighting arrived at the Dniepr near Shklov. There the SS units ran into the Soviet 13th Army, which had launched a counterattack in the direction of Mogilev-Smolensk. This marked the beginning of the Battle of Smolensk, which was to cost both sides heavy losses.

At the end of July *Panzergruppe Guderian* was in danger of being encircled by the Russians. At heavy cost, the *SS-Division "Das Reich"* defended the salient northwest of Yelnya. When the battle ended in mid-August and the German troops returned to the offensive, *Panzergruppe Guderian* was turned south to take part in the Battle of Kiev. The division was positioned on the *Panzergruppe*'s right flank and advanced west of Romny south into the area east of Prikuli.

While the battle was still going on, *"Das Reich"* was transferred to the newly-inserted Pz.Gruppe 4 in preparation for "Operation Typhoon", the assault on Moscow. The attack on the Soviet capital began on 2 Octo ber 1941. Progress was initially good, however the onset of bad weather with its associated mud and later snow and ice, slowed the offensive.

A report from those days stated:

> "The battalions of the SS-Division *"Das Reich"* worked their way forward, far ahead of all other divisions. In spite of the unfavorable weather the division continued to advance and took Gzhatsk.

There it was halted temporarily and for six days was engaged against two entire Russian armies!"

From then on the weather limited all movement. First there was mud, in which the vehicles became bogged down; then came snow and ice, and the troops could only advance on foot. Not until 18 November did the *4. Pz.Armee* launch the final assault on Moscow between Volokolamsk and Naro Fominsk. Eight days later the division reached the Istra River near Moscow. The temperature was minus 35 degrees! The advance continued along the Volokolamsk road. In spite of fierce resistance the *Regiment "Deutschland"* advanced along the Rsachdestwo-Lenino road. The commander of the regiment, Ob.Sturmb.F. Kumm, wrote:

> "The men of the regiment believe that they can count on the fingers of one hand the days until they enter Moscow. In the clear, cold weather the spires of the city can be seen without binoculars."

However, the advance went no farther. On 6 December 1941 the Soviets launched a counterattack against Army Group Center. Equipped for winter conditions, the superior forces of the Red Army overran the German front lines and drove back the army units far to the west. Then in early 1942 the *SS-Division "Das Reich"* was pulled out of Army Group Center's front and transferred to the Protectorate to rest and reequip. Virtually every regiment, *Standarte*, battalion and *Sturmbann* had lost two-thirds of its original complement of personnel.

Following reformation of the division in the summer of 1942 at troop training grounds in Bergen and Fallingbostel and the necessary training and unit exercises, it went to northern France as an occupation unit. There the division came under the command of the AOK 15 and on 9 November 1942 was renamed SS-Pz.Gren.D. *"Das Reich"*.

The division now included the *Pz.Gren.Standarten "Deutschland"*, *"Der Führer"* and *"Langemarck"*, the *Pz.Rgt. "Das Reich"*, plus other units associated with an army panzer-grenadier division. The renamed division remained in northern France as an occupation unit under the command of the *15. Armee*. The *Standarte "Langemarck"* was disbanded in the summer of 1943.

In February 1943 the division was transported to Russia, where it joined Army Group South. The division was attached to the II. SS-Pz.K.

and took part in the recapture of Kharkov. In July 1943 *"Das Reich"* was committed with the *"Leibstandarte"* and *"Totenkopf"* Divisions in "Operation Zitadelle". The division fought well but suffered heavy losses, especially in tanks.

There followed a brief period of rest in the rear area of Army Group South and in August the division returned to action with the *4. Pz.Armee* between Stalino and Poltava. The renumbering of all units of the *Waffen-SS* took place during the general retreat to the Dniepr.

On 22 October 1943 the division received its ultimate designation: 2. SS-PD. *"Das Reich"*.

In the autumn the division took part in Army Group South's withdrawal through Kiev and Zhitomir until the encirclement of the *1. Pz.Armee* around Kamenets-Podolsk. After a successful breakout at Vinnitsa, in March 1944 the division was transferred to the area of Toulouse in southern France to rest and reequip. A mixed battle group from the division under the command of Headquarters, SS-Pz.Gren.R. 3 remained on the Eastern Front for several more weeks.

After the start of the invasion the division was hastily transferred to Normandy. There the division took part in fierce defensive fighting and was encircled at Falaise with the 7. *Armee*. After breaking out of the encirclement the division was transferred to the rear to rest and reequip. In October 1944 *"Das Reich"* was transferred into the area around Paderborn.

The next major action for the division came during the Ardennes offensive which began on 16 December 1944. The SS-PD. *"Das Reich"* was in the second line at the start of the offensive, but it was soon inserted into the left wing of the *6. SS-Pz.Armee* after the infantry divisions had suffered heavy casualties. The snow-covered hill country of the Ardennes proved to be a serious obstacle and the division suffered heavy losses. In spite of this it was able to advance approximately 25 kilometers to the west by Christmas Day 1944.

Then, however, the skies cleared and the Allied air forces intervened. Every subsequent attack was stifled by a hail of bombs. Like all of the other units, *"Das Reich"* was forced to retire to its starting positions in the Eifel. From there the entire division was transferred to Hungary in January 1945. In February an attack was launched with the objective of relieving Budapest, however it was unsuccessful.

The Red Army pursued the retreating German forces. The division retreated to the northwest and on 8 May 1945 surrendered in the Linz area of Austria.

Commanders:
O.Gruf. Hausser; Ob.F. Bittrich; Brig.F. Kleinheisterkamp; Brig.F. Keppler; Gruf. Krüger, W.; Gruf. Ostendorff; Ob.Sturmb.F. Tychsen; Gruf. Lammerding; Ob.F. Baum; Stand.F. Lehmann; Brig.F. Vahl.

3. SS-Panzerdivision "Totenkopf"

With the agreement of the OKH, in October 1939 the *Reichsführer-SS* ordered the formation of a division from the *"Totenkopfverbänden"* (Death's Head Units) stationed in the Reich. Thus the division received the name which it retained until the end of the war. After the necessary basic training and unit exercises had been completed, plus instruction of officers and non-commissioned officers, in March 1940 the division was deemed fit for service and was transferred to western Germany as OKH reserve.

With the start of the western campaign on 10 May 1940, the division was sent to Pz.Gruppe Hoth. There it was attached to the XXXIX. AK.mot. and deployed in the front line on the Belgian-French border. Beginning on 20 May the division fought north of Arras and later southwest of Lille.

The second phase of the campaign saw the division advance into southeastern France as part of Pz.Gruppe Guderian. The unit reached the area around Lyons. From there it was transferred to Pz.Gruppe von Kleist and moved quickly to the Spanish border, which it reached near Irun on 25 June 1940.

After being transferred back to the homeland the unit was reorganized as a motorized division. This conversion was completed in April 1941 and in May 1941 the *"Totenkopf"* D.mot. was transferred into the area around Tilsit, East Prussia where it was attached to Pz.Gruppe 4 as a reserve unit.

In the first days of the war against the Soviet Union the *Panzergruppe* moved the division into the Dvinsk area to secure the bridgehead there. The division suffered extremely heavy losses in its first days in action, in particular as a result of skilled attacks by light Russian tanks. The losses forced the division to temporarily disband one regiment.

The fighting in the Dvinsk bridgehead lasted until the end of the first week of July, by which time *Panzergruppe 4* had cleared its way. The *SS-Division "Totenkopf"* was moved to the right wing and it laboriously fought its way through Sebesch to Dno. There it was on the extreme right flank of Army Group North and had to struggle to maintain contact with the division on the left wing of Army Group Center.

At that time the 8. PD. was temporarily encircled by the enemy and the division was called upon to dispatch a regiment-strength battle group to free the panzer unit. At the end of the month the German advance halted temporarily at the Mga River line. The division gained ground around Schimsk. Not until mid-August could the advance be resumed. It took the *"Totenkopf"* Division four days to cover the 260 kilometers from the area near Luga to Kholm southwest of Lake Ilmen. From there, in the last days of August the division reached the battlefield where it was to remain for the next few months: Demyansk.

When the year 1941 ended the division was located approximately 40 kilometers southeast of Lake Ilmen on the Waldai Hills. The hard, cold winter forced the unit to fight for shot-up village houses and frozen infantry trenches. The division was still there when Soviet units succeeded in encircling the entire German II. AK. in February 1942. The *"Totenkopf"* Division played a major role in "Operation *Fallreep*", whose objective was to restore contact with the German front. On 19 April 1942 assault groups from the division crossed the Lowatj and established contact with the 5. Jäg.D. approaching from the west. The "Demyansk Pocket" had rejoined Army Group North; from that time on that sector of the front bore the name "land bridge".

In the summer of 1942 the *"Totenkopf"* Division was pulled out of Army Group North's sector and transferred to southern France for reorganization as a panzer-grenadier division. In the first weeks of 1943 transport trains brought the reequipped division to the southern sector of the Eastern Front. There the regiments were attached to the II. SS-Pz.K. With the rest of the corps the division took part in the recapture of Kharkov.

By April the division was concentrated in the area northeast of Poltava where it was readied for "Operation Zitadelle". The corps was subsequently transferred to the right wing of the 4. Pz.Armee. The division, with a strength of 104 combat vehicles, was supposed to maintain contact with *Armee-Abt. Kempf* (168 ID.) on the Pz.Armee's right.

When the attack began on 5 July 1943 the attackers were able to breach the first enemy positions, however the division's tanks were halted in front of the second line. Not until the second day were these positions breached, however the unit on the division's right did not keep pace. On the evening of that day the division's Panther tanks were 18 kilometers from their starting position.

On the third day of the battle powerful forces of the Soviet 5th Guards Tank Army and the 6th Army appeared on the battlefield and hampered the German attack. Rain also hindered the advance, which by 11 July reached Prokhorovka. Then approximately 1,000 Soviet combat vehicles rumbled to meet the II. SS and III.Pz.K. The next day the wrecks of 220 Russian and 150 German tanks littered the battlefield.

The German advance was halted. The next day the II. SS-Pz.K. – without the Division *"Totenkopf"* – was pulled out of the front and transferred to northern Italy to disarm the Italian Army.

In August 1943 the division was transferred into the Stalino area. From there it fell back to the Dniepr, which was defended northwest of Dnepropetrovsk until October. Powerful Soviet units crossed the river in October resulting in weeks of fighting for Krivoy Rog, with the division alternating between attack and defense as part of the LVII. Pz.K. of the 1. Pz.Armee.

On 22 October 1943 the units of the *Waffen-SS* were numbered and the division received its ultimate designation: 3. SS-PD. *"Totenkopf"*.

The new year 1944 saw the panzer division involved in continuous defensive movements through the area around Kishinev and Cherkassy. At the end of June 1944 a major Soviet offensive resulted in the collapse of Army Group Center. In July the 3. SS-PD. was transferred to Army Group Center, which committed it to the defense of Bialystock and from August 1944 of Modlin. The division remained in defensive positions there until the end of the year.

After the high command of the 6. SS-Pz.Armee was transferred from the Western Front to Hungary to relieve Budapest, the 3. SS-PD. was relieved from its positions in and around Modlin and followed the SS panzer army headquarters to Hungary.

There the division was attached to the IV. SS-Pz.K. On 18 January 1945 the corps (from right to left: 3. PD., 5. and 3. SS-PD.) attacked to the northeast from south of Stuhlweißenburg. The 3. SS-PD. advanced through Seregelyes to Adony on the Danube south of Budapest. It advanced no farther, however; on 27 January 1945 the battle was broke off and Budapest was left to its fate.

In March 1945 the 6. SS-Pz.Armee began a retreat which took the division through the Vienna area to Linz in Upper Austria. There the story of the 3. SS-PD. ended.

Commanders:
O.Gruf. Eicke; Brig.F. Prieß; Brig.F. Becker; Ob.F. Deisenhofer.

4. SS-Polizei-Panzergrenadierdivision

Following the outbreak of the Second World War the *Reichsführer-SS* and the chief of German police ordered the formation of a division from officers, non-commissioned officers and enlisted men from the *Ordnungspolizei* (uniformed regular police) and the *Schutzpolizei* (municipal police). Formed in southwestern Germany in October 1939, the unit remained a part of the police force. Its members therefore continued to wear police uniforms and use police ranks. The OKH (especially *Wehrkreis III*, Berlin) had to detach the AR. 300 and divisional units such as the pioneer battalion, signals battalion, etc, all with the number 300, to the division. The new unit was dubbed the *Polizei-Division.*

After necessary training the division was transferred to the Upper Rhine front and was attached to the 7. *Armee*. The positional warfare on the Upper Rhine in the first months of the war made no military demands of the division. During the winter of 1939-40 the division was withdrawn from its positions in the *Westwall* and was transferred to the rear for further training in the area Tübingen-Reutlingen-Nürtingen.

In mid-May, with the western campaign in its initial phase, the division was placed on alert and transferred to Luxembourg and southeastern France. There it was attached to the XXV. AK. of the *12. Armee*. The army was preparing for the second phase of the offensive. When the attack began on 9 June the Pol.D. was initially held in reserve, however it was soon inserted into the front line next to the 10. and 26. ID. After five days on the attack the division approached the fortress area of Verdun from the northwest. When the area was taken by the neighboring corps, the Pol.D. turned south to pursue the fleeing enemy. This was continued until the halt order was issued.

When the eastern campaign began the division was still in the west, but in July it was transported by train to Army Group North. At the end of July it was sent to the LVI. AK.mot. in the Luga area. The corps had orders to drive to Leningrad with other units of Pz.Gruppe 4.

Advancing through jungle-like forest, the division gained ground, but slowly. At the end of August the Soviet 2nd Militia Division was smashed and in the first weeks of September the division fought its way toward Krasnogvardeysk. The Pol.D. suffered heavy casualties, especially during the taking of Luga. The division was attached to the L. AK., whose two divisions had to attack via Krasnogvardeysk directly toward Leningrad. After the fall of the transportation center the Pol.D. attacked Pushkin, formerly Tsarskoye Selo, grand residence of the former tsars. Supported by a battle group from the 1. PD., the SS-Pol.R. 2 took the former imperial residence. It was Army Group North's last success before Leningrad.

In mid-October the weather turned and four weeks later the land around Leningrad became frozen in snow and ice. Prosecuting the war became impossible. The L. AK. received orders to seal of the city from the south. One of the divisions under its command was the SS-Pol.D. It held a front just south of the metropolis, which the policemen could easily observe with binoculars.

With temperatures dropping to minus 40 degrees, positional warfare was the only possibility.

The division's next area of operations was the region of swamps and forests west of the Volkhov. In February 1942 the Soviets crossed the Volkhov and advanced to Lyuban. The division was pulled out of the Leningrad front and rushed to the threatened sector. In the German counterattack aimed at sealing off the pocket front, the Pol.D. attacked from Spaskaya Polisty south in the direction of Mostki. It linked up with German battle groups attacking from the south and sealed the pocket around General Vlasov's army. The Pol.D. was subsequently transferred back to the Leningrad front and at the beginning of January 1943 was southeast of the city on the Neva.

This was the Pol.D.'s last area of operations in the northern sector of the Eastern Front, for in September 1942 at the SS Troop Training Grounds at Heidelager (Debica) near Krakow formation began of a new division which was to replace the Pol.D.

First two police panzer-grenadier regiments (Pol.Pz.Gren.Rgter 1 and 2), each of three battalions, were formed from former occupation troops in Poland. Then were added an artillery regiment with six batteries, a reconnaissance battalion, flak battalion, assault gun battalion and signals battalion. On 1 June 1943 these units were designated Pol.Pz.Gren.D. and on 22 October the division received its ultimate title, 4. SS-Pol.Pz.Gren.D.

In December 1943 the units of the new division were transported to Army Group E in the Saloniki area. The 4. SS-Pol.Pz.Gren.D. remained their under the command of the *Oberfeldkommandantur 395* (Military Government Area Headquarters). The elements of the former Pol.D. still in northern Russia were used up in Army Group North's retreat from the Leningrad area to the Luga. The survivors were moved back to Pskov to rest and reequip and in April 1944 were transferred to Greece.

The police uniforms previously worn by the Pol.D. were now replaced with the uniforms and rank badges of the *Waffen-SS*.

In August 1944 the now complete 4. SS-Pol.Pz.Gren.D. was transferred into the Belgrade area and began anti-partisan operations in the Skolpje area and in the Banat. One month later it began operations against forces of the Red Army advancing toward Hungary. At various times the division was attached to the IV. and LVII. AK. of the *6.* and *8. Armee*, respectively. There was mixed fighting during this deployment, especially in the Szolnok bridgehead.

In January 1945 the badly battered division was withdrawn from the Hungarian front and transferred to Pomerania to rest and reequip. There was little time, for the Soviet had already advanced to Pomerania and the Baltic Coast. While some elements of the division saw action around Burtin, Stolp and Konitz, other units fought at Gotenhafen, the Oxthöfer Kempe and near Hela. In mid-April the remnants of the division were shipped to Swinemünde, from where they were committed around Eberswalde, Oranienburg and Neubrandenburg under the 23. SS-Gren.-D.

Commanders:

Gruf. Pfeffer-Wildenbruch; Brig.F. Mülverstedt; Brig.F. Krüger, W.; Brig.F. Wünnenberg; Ob.F. Schmedes; Brig.F. Vahl; Brig.F. Schmedes; Stand.F. Harzer; Ob.F. Dörner.

5. SS-Panzerdivision "Wiking"

While the western campaign was still in full swing, in May-June 1940 hundreds of young men from Denmark, Norway, Holland and Belgium volunteered for service in the German armed services. In the autumn and winter of 1940-41 they were incorporated into the newly-formed SS regiments *"Nordland"* and *"Westland"*. When the flow of volunteers showed no signs of subsiding, the SS command decided to expand these units and form a volunteer division which received the name *"Wiking"*.

After just a few weeks the new division was transported to southeastern Poland. The division detrained in the Lublin area and was attached to Pz.Gruppe 1. *"Wiking"* was in the forefront of the German attack on the Soviet Union from the first day. On the second day of the attack German forces broke through the Soviet border positions and began a rapid advance into the interior.

Attached to the XIV. AK., by the eighth day of the campaign the *"Wiking"* Division had reached a position south of Zhitomir. The advance continued to the southeast and by the end of the first month of fighting the German motorized units had reached the Taracha area, just west of the mighty Dniepr River. There the division had to turn south to help secure the German encirclement front around Uman to the southwest and west to the Dniepr.

Then the division resumed its advance to the Dniepr, now attached to the III. AK.mot. In the first week of August the division took Kremenchug and exactly eight days later Dnepropetrovsk. The division did not take part in the battle of encirclement at Kiev as the III. AK.mot. (redesignated as a panzer corps on 3 October) was ordered to attack toward the Sea of Azov. This objective was achieved and the advance continued to the Crimea, but then the Russian winter set in. The unit was forced to withdraw to the Mius front and go over to static warfare in improvised trench positions.

The *"Wiking"* Division was pulled out of the line. Rested and reorganized as a fully-motorized division, in spring 1942 *"Wiking"* was ready for the German summer offensive. It was part of the *1. Pz.Armee* whose objective was to storm the Caucasus with the *17. Armee*.

On 21 July 1942 the division crossed the Donets and advanced in the direction of Rostov. Three days later the division's vehicles rolled through the city and crossed the broad Don to the south. Before them the German and European volunteers found endless steppes, a type of terrain not seen in central Europe.

The division crossed the Kuban near Kropotkin and advanced through Armavir south to the oil region of Maykop. The soldiers were now in the foothills of the Caucasus mountains. At the end of August the corps on the right took the significant Black Sea port of Novorossisk. The *Armee* subsequently ordered the LVII. Pz.K. and the XXXXIV. AK. to advance through the Caucasus and take Tuapse.

At the end of August the attack bogged down in back-and-forth fighting for possession of mountain villages, narrow roads, steep mountains and high stony plains. Crossing the mountains, which were covered in ice and snow from October on, proved impossible. The German advance on Tuapse was called off in late autumn; the troops dug in.

In October 1942, during the defensive struggle in the Caucasus, the division was renamed the 5. SS-Pz.Gren.D. *"Wiking"*.

The division was transferred east. Together with the 13. PD. it was to take the city of Grozny at the beginning of October. The division history states:

> "*SS-Gruppenführer* Steiner knew, however, that the Soviet barricade blocking the road from Grozny to the Caspian Sea had to be eliminated. He therefore divided the division into the four columns. *"Nordland"* was to advance on either side of the Kurp River toward Malgobek. The *Panzerabteilung* received orders to force a bridgehead. In cooperation with it *"Westland"* was to eliminate the enemy in Sagopshin. The pioneers were also to advance along the Kurp…"

That was the plan; but the enemy was stronger. When the division's attack went ahead at 0500 hours on 26 September 1942 the enemy struck back. His massed artillery and infantry in well fortified mountain positions halted the division's advance after initial success.

The attack by the division was called off and instead it was ordered to maintain contact with the LII. AK. near Mozdok. This was the end of the advance, for in January 1943 the Soviet army went on the offensive. The *6. Armee* was encircled at Stalingrad and the Caucasus became a secondary front.

The subsequent retreat took *"Wiking"* across the Manych northwest into the area of Izyum and Kharkov. There the division was attached to the XXIV. Pz.K. of the *1. Pz.Armee*. In April 1943 the division was *Armee* reserve southwest of Slavyansk. By that time it had just 31 tanks and six assault guns.

Following the failure of "Operation Zitadelle" the Soviets launched a major offensive aimed at destroying the German army groups in the south. The division was forced to withdraw to the Dniepr with the *8. Armee*.

In January 1944 the entire *Armee* was hit by the 2nd Ukrainian Front's massive offensive between Nikopol, Cherkassy and Uman. Then the armored forces of the 1st Ukrainian Front also attacked. The two armored spearheads linked up, trapping the bulk of the *8. Armee* in the Korsun area. In the pocket with the 5. SS-PD *"Wiking"* (the unit had been designated a panzer division on 22 October 1943) were four complete infantry divisions, elements of three more and the *SS-Brigade "Wallonien"*. Hitler forbade a breakout. Not until all counterattacks from outside failed was the necessary order issued on 15 February 1944.

"Wiking" was on the left wing of the breakout force. In ice-cold winter weather it attacked enemy positions near Komarovka, broke through to the southwest in the direction of Pochapintsy and split up into two battle groups. These worked their way over the snow-covered hill terrain to Lysyanka where they met the spearheads of the 1. PD. approaching from the west.

Once again the badly battered division had to help free surrounded German units. The Red Army's big offensive had reached Kovel, where various elements, including rear-echelon units, were trapped. The 5. SS-PD. was sent in from the north. Under the command of their new division commander Gruf. Gille, the SS men were able to liberate Kovel. For this feat the division commander became the first member of the *Waffen-SS* to receive the highest German decoration, the Knight's Cross with Oak Leaves, Swords and Diamonds.

Between May and July 1944 the division was ordered back to the Debica troop training grounds to reequip with new weapons, materiel and vehicles. By August 1944 it was back at the front, this time as part of the defense of Modlin. It remained there under the IV. SS-K. until the end of the year.

At the beginning of 1945 it was hastily transferred to Hungary along with many panzer divisions. At the beginning of January the *6. SS-Pz.Armee* was supposed to liberate Budapest. The last German offensive of the war began on 18 January 1945. *"Wiking"* was positioned between the army's 3. PD. and the 3. SS-PD *"Totenkopf"* on the left. The German attack surprised the Russians. The regiments and battalions of the *"Wiking"* Division overcame the initial enemy resistance north of Lake Balaton, advanced east, captured the town of Sarosd and advanced northeast toward Adony. There the attack spearheads linked up with those of the 3. PD. But *"Wiking"* went no farther. On 27 January 1945 the attack was halted.

The Red Army now launched a big offensive. The German units withdrew north of Lake Balaton and reached Austria. The battle groups of the 5. SS-PD. retreated to the northwest and were near Graz when the surrender came.

Commanders:
O.Gruf. Steiner; Gruf. Gille; Ob.F. von Scholz; Stand.F. Mühlenkamp; Stand.F. Ullrich.

6. SS-Gebirgsdivision "Nord"

Formation of this division prior to the start of the eastern campaign marked the end of the "second wave" in the development history of the combat divisions of the *Waffen-SS*. All other major formations of the *Waffen-SS* were created after the invasion of the Soviet Union and the beginning of anti-partisan warfare in the Balkans and finally after the Allied invasion of France.

Following the completion of the Norwegian campaign in June 1940, the *Waffen-SS* formed the so-called *"SS-Kampfgruppe Nord"* (SS Battle Group North) from Upper Bavarian and Austrian SS units for the purpose of securing the mountainous country. Two SS mountain regiments were formed with several supply companies.

After the beginning of the eastern campaign *SS-Kampfgruppe "Nord"* was expanded to brigade size and in August 1941 was transferred to central Finland. Later it was moved to northern Finland. Brig. *"Nord"* participated in the German advance on Aalakurtti. Later it took part in the fighting associated with the advance across the Tuntskoy River in the direction of Kandalaksha and between the Voyta River and Lysaya.

When the 1941 campaigning season came to an end, so too did the advance of the German troops in Finland. The troops had to go over to the defensive in the frozen ground of northern Europe. The last weeks of the winter were used to expand the brigade into the *SS-Division "Nord"*.

When the sectors at the front were reorganized between German and Finnish forces in the summer of 1942, the *SS-D. "Nord"* and the newly-arrived 7. Geb.D. belonged to the XVIII. Geb.K. of the *20. Geb.Armee*. The corps had taken over the area of operations formerly held by the Finnish III. AK. in the Louhi area.

The division's face had changed in the Arctic summer months. It had been transformed into a mountain division and in the summer 1942 consisted of two mountain regiments, a mountain artillery regiment of four battalions, a motorized rifle battalion, a pioneer battalion, a reconnaissance battalion, an anti-aircraft battalion and a signals battalion plus supply units. The missing anti-tank battalion was created at the Wildflecken troop training grounds from the *Heeres-Pz.Jäg.Ers.Abt. Hilversum* (army anti-tank replacement battalion).

The division remained in the sector west of Louhi, the so-called "Kiestinki Sector". On 22 October 1942 it received the official designation *6. SS-Geb.D. "Nord"* and its two mountain regiments were numbered 11 and 12.

In the summer of 1944 Finland dropped out of the war against the Soviet Union and its units began fighting alongside the Red Army. In September 1944 the Germans launched "Operation *Birke*" (Birch), the withdrawal of their forces from Finland.

At the beginning of September 1944 the *SS-Geb.D. "Nord"* abandoned the positions it had held for a year in the Kiestinki front. Constantly engaged by Russian and Finnish units, it withdrew via Kuusamo and Rovanniemi to Munio in the Karasuanto position. The division's rear guards often had to fight tough and costly engagements against the enemy's motorized troops, which repeatedly outflanked and encircled the German columns.

The German forces in Finland were able to reach the old Finnish-Norwegian border in relatively good order. The *SS-Geb.D. "Nord"* marched via Skibodden to Narvik from where it was transported by rail to Oslo. There the SS men went aboard ships which transported them to Jutland by mid-December 1944. There the unit rested, took in replacements and reequipped with new weapons. Then it was transported by rail through all of western Germany. Not until they reached the Pirmasens area did the *SS-Gebirgsjäger* leave the trains.

The division was attached to the *1. Armee* (LXXXX. AK.). The *Gebirgsjäger* were practically thrown from the railway coaches straight into the battle against the American forces advancing toward the Rhine and the Moselle.

The battle groups of the *SS-Geb.D. "Nord"* took part in the fighting at Wingen, Wimmenau, Melch, Baerenthal and between Bitsch and Rothbach. In February 1945 it conducted a fighting retreat in the Reipertsweiler and Hanauerweiher area. The next month brought a transfer into the area southeast of Trier followed by fighting at Simmern on the Moselle, near Brodenbach and in the Osburger Hochwald. Several times the division was dispersed and its battle groups fought alone.

Then followed the retreat across the Rhine south of Boppard. On 1 April 1945 the division numbered 20,000 men with 600 horses and mules, 350 motor vehicles, 6 mountain cannon (without ammunition), 3 anti-tank guns, 1 assault gun, 2 captured American tanks and 200 American prisoners of war.

The division was inserted into the front in the Limburg/Lahn area. There followed defensive actions and withdrawal movements through Usingen and Assenheim. The division was split up into battle groups which fought their last engagements in and around Lindheim, Waldensburg and Leisenwald. When the surrender took effect the last survivors of the *SS-Geb.D. "Nord"* were in the Büdinger Forest in the Taunus Mountains.

The very last belonged to the former *SS-Schi-Btl. 506* (SS Ski Battalion). They were Norwegian volunteers who had been attached to the division since September 1943. The battalion reached the Nuremberg-Bayreuth autobahn and laid down its weapons in Reit im Winkel on the day of the surrender.

Commanders:
Stand.F. Hermann, Stand.F. Demelhuber; Stand.F. Schneider; Brig.F. Kleinheisterkamp; Stand.F. Debes; O.Gruf. Krüger, F.W.; Gruf. Brenner; Stand.F. Schreiber.

7. SS-Gebirgsdivision "Prinz Eugen"

This division was the first to include in its ranks ethnic Germans living in eastern and southeastern Europe and later non-German volunteers from these occupied territories.

Beginning in February 1942, the office of the senior SS and police commander in southeast Europe began recruiting ethnic Germans from the greater Serbian region for a new SS division. Formation of the unit, which included two mountain regiments, took place in October 1942 in the Serbian Banat. The division was given the name *"Prinz Eugen"* and on 1 November 1942 was placed under the commander of German troops in Serbia.

On that day the division was one of only two large formations based in the Balkans. The other, the 22. ID., was stationed on Crete. Then, on 17 December 1942, Hitler issued the order "...to smash the partisan movement from the ground up!"

The execution of this order took the form of "Operation White", which began on 20 January 1943. An extensive pincer movement was supposed to clear southern Croatia and Bosnia where Tito's partisans had become extremely active, virtually paralyzing German-Italian supply traffic.

The *"Prinz Eugen"* Division attacked from the Karlovac-Prijedor-Banja Luka area. After nine days of fierce fighting the division stormed the heavily fortified town of Bihac. Three days later the 3rd Banija Partisan Division was smashed southeast of Bihac and Bosnian-Petrovac was taken. The pursuit continued into the mountain country and at the end of February the mountain troops of the *"Prinz Eugen"* Division stormed Bosanso-Grahowo and drove the 1st Bosnian Partisan Corps back into the Sator Mountains.

After several weeks of static warfare and the receipt of replacement personnel and equipment, the division took part in "Operation Black". The latter began on 15 May 1943 and its objective was to trap approximately 20,000 partisans under the personal command of Tito. The terrain in which the offensive was conducted was described as follows:

> "Mostly deep, canyon-like gorges through which wild, barely passable rivers roared; between them sparsely overgrown plateaus approximately 50 kilometers long and 30 kilometers wide..."

The battle in the mountains of Montenegro was a success, however Tito himself was able to escape with about 3,000 men. The division remained in action for months, placing great physical demands on the troops. In September 1943 it was transferred to the area around Split to defend the coastline. Then in October 1943 the division received its ultimate designation: *7. SS-Geb.D. "Prinz Eugen"*. Its two mountain regiments were numbered 13 and 14.

In December 1943 the division was transferred again, this time to Sarajevo for occupation duty. One month later, however, it returned to action against the partisans in Croatia. For the next seven months the division fought an almost hopeless battle against Tito's partisan units which, thanks to British and Soviet support, were growing stronger by the day.

In August 1944 the division was rushed to Serbia. Soviet forces had crossed the borders of the former Kingdom of Yugoslavia and the Romanian army had gone over to the enemy. The division now faced a new foe, the Red Army.

Until December 1944 the division fought as part of the V. SS-K. of the *2. Pz.Armee*. In November and December of 1944 it suffered heavy losses in men and materiel in weeks of fighting in and around Nisch.

The year 1945 brought the Red Army's unstoppable advance into Bulgaria and Yugoslavia and the evacuation of Greece by German forces. The 7. SS-Geb.D. joined the general retreat by Army Group E. In turn it was attached to the XXXXIV., LXXXXI., XXI. And LXIX. AK.

The retreat took the division through its former combat zone of Croatia, which was home to many members of the division. In March and April 1945 the unit reached Agram and continued to move north. The Zvonimir position was abandoned at the end of April. The division reached the Drau Valley, passed Cilli and the Mies Valley and crossed the southeast border of Carinthia. Then the war ended.

This marked the beginning of a period of captivity in the hands of the Tito partisans from which only a very few returned.

Commanders:
O.Gruf. Phleps; Brig.F. von Overkamp; Brig.F. Kumm; Ob.F. Schmidhuber.

8. SS-Kavalleriedivision "Florian Geyer"

The division was the second and last large unit of the *Waffen-SS* formed in 1942. An SS cavalry brigade, also known as *SS-Kav.Brig. "Fegelein"*, had existed since 1940.

Shortly after the outbreak of war SS-Stand.F. Fegelein had been named chief cavalry officer of the *Allgemeine-SS*. From the existing cavalry formations of the SS he created two cavalry *Standarten* which it was thought would serve as reconnaissance troops for the existing divisions of the *Waffen-SS*. When these division units were made fully motorized, however, this notion was dropped. In February 1941 the two *Standarten*, which were stationed in Warsaw and Lublin, were renamed SS-Kav.R. 1 and 2.

By October 1941 these two regiments had been expanded into a brigade through the addition of pioneer, signals, flak and supply units. The cavalry regiments were already in action, however, having been committed behind Army Group Center in the Pripyat Marsh region soon after the start of the eastern campaign. Operating in an area unsuited to motor vehicles, the cavalrymen searched out Soviet battle groups and the first partisan units.

The Kav.Brig.'s great test came in the winter of 1941-42 when the cavalry, the only troops to retain their mobility in the ice and snow, were inserted into the front to seal gaps between the infantry divisions. The brigade played a major role in preserving the integrity of the front in the Rzhev area.

In spring 1942 the Kav.Br. *"Fegelein"* was pulled out of the front and transported to the troop training grounds at Debica near Krakow. By this time it was down to a few hundred men. On 21 June 1942 the brigade became the SS-Kav.D., which was superior to the army's disbanded 1. Kav.D. in organization and equipment.

The division consisted of three cavalry regiments (which were later numbered 15, 16 and 17), bicycle, anti-tank, flak and signals battalions, pioneer and trained replacement battalions, the AR.(mot.) 18, and supply troops. When the units of the *Waffen-SS* were numbered the division received the designation *8. SS-Kav.D. "Florian Geyer"* and the attached units other than the cavalry regiments were assigned the number 8.

Soon after the division's formation it was sent to the front. Among the cavalrymen were many ethnic Germans from Hungary, men who had lived on the country's big farms and estates.

In October 1942 the 8. Kav.D. was employed in Army Group Center's rear in the Smolensk area. In November it went to the sector around Rzhev, where Soviet armored forces had cut the Rzhev-Sychevka rail line. The 8. SS-Kav.D. moved into the area north of Bely. On 7 December 1942, with the snow 40 cm deep, the division, together with the 19. and 20. PD., attacked forward elements of the Soviet 30th Army. The attack force linked up with units advancing from the south and sealed the pocket around the enemy army. The Battle of Bely lasted six days.

When German forces retreated from the Rzhev salient in March 1943, the division was moved to the Orel area as reserve to the *2. Pz.Armee*. In April it was moved to Bobruisk where the cavalry division rested and reequipped. July and August 1943 saw the division in constant action against partisan units in Army Group Center's rear, with the cavalry troops again operating in the Pripyat Marshes.

September 1943 brought the 8. SS-Kav.D. back to the *8. Armee*'s front during fighting west of Kharkov. After a transfer to the neighboring *1. Pz.Armee* – likewise west of the Dniepr – the division took part in the *Armee*'s fighting withdrawal until December 1943.

In 1944 the cavalry division was sent to Hungary as an occupation unit. There its units were reorganized and the division took in replacements and new weapons. The restoration of the division was completed in the Budapest area by the end of March. From there the division was transferred to Esseg in Croatia, occasionally engaging partisan groups of Tito's army.

In August 1944 the division was relieved and transferred to the front in Transylvania. There it fought Russian forces advancing from Romania into Hungary. In November 1944 it joined the withdrawal to Budapest under the command of the IX. SS-Geb.K.

On Christmas Eve 1944 the Hungarian capital was encircled by divisions from three Soviet armies. With most of the horses gone, the men of the division fought as grenadiers, defending the eastern quarter of the city. In January 1945 German forces evacuated that part of the city and withdrew across the Danube to Buda. On 11 February 1945, with ammunition almost gone and the counterattack by German armored forces halted far from the city, the city's defenders surrendered.

Thus ended the history of the *8. SS-Kav.D. "Florian Geyer"*.

Commanders:

Gru.F. Fegelein; Brig.F. Bittrich; Stand.F. Lombard; Brig.F. Rumohr.

9. SS-Panzerdivision "Hohenstaufen"

This division was established on 1 February 1943 at the former cadet institute in Berlin-Lichterfelde. Initially it consisted of personnel from the *"Leibstandarte"* replacement units, however volunteers from all parts of the Reich were recruited to bring the unit up to full strength. Soon after formation of the operations staff, the units moved to Mailly-le-Camp near Reims and in April 1943 to Ypres. There the new panzer division received its name *"Hohenstaufen"*. The division was organized like the existing panzer divisions of the *Waffen-SS*. The panzer regiment and the other division units received the number 9; the panzer-grenadier regiments were given the numbers 19 and 20.

After establishment was completed the panzer division was attached to the *Heeresgruppenkommando "D"* as reserve and in March 1944 was placed at the disposal of the AOK 19 in southern France.

The division remained in France just a few days, and in April 1944 it was rushed to the *4. Pz.Armee* in southern Russia. The units of the division left the trains in the Ternopol area and were immediately sent to relieve the German units encircled near Kamenets-Podolsk. The division's deployment at the front in the east lasted just three months, for in the west the Allies had invaded France. In June the division was rushed west to join the II. SS-Pz.K.

The corps headquarters, which had just arrived from Poland, had under its command the advance detachments of the 9. and 10. SS-PD. These were immediately committed to a counterattack against the British VIII Corps. On 28 June 1944 the British had launched an attack from Tilly against Caen. Elements of the division drove the British from the hills around Gavrus and prevented the loss of Caen.

At the end of July-beginning of August American forces breached the German front near Avranches and drove deep into western and central France. The division was moved into the Mortain area to participate in a German counterattack. The attack was a failure, and in mid-August the Allies encircled the entire *7. Armee*, including the II. SS-Pz.K., at Falaise. The 9. SS-PD. took part in difficult, costly fighting between Mortain and Domfront. The SS troops broke through the enemy lines at great cost and then took part in a counterattack to help other German units escape the encirclement.

At the end of August the 9. SS-PD. was transferred to the area around Arnhem to rest and reequip. It remained part of the II. SS-Pz.K. and until

16 September 1944 was a reserve unit of Army Group B. That was the day the huge British and Canadian airborne operation began in the Arnhem area.

When the operation began, *Feldmarschall* Model, the Commander-in-Chief of the army group, was at the command post of the II. SS-Pz.K. Immediately after receiving initial reports on the airborne landing he committed the first available battle group from the 9. SS-PD. under Sturm.F. Spindler and ordered the division to:

> "…assemble in the Velp area and stand ready to attack the enemy forces which have landed near Oosterbeek west of Arnhem."

The division's Panther tanks were still on trains somewhere in western Germany; however, under the command of Ob.Sturmb.F. Harzer, on 20 September the men of the division, together with *Luftwaffe* troops and sailors from the replacement units, attacked the British forces which had landed near Arnhem. They systematically eliminated nests of resistance and on 22 September Arnhem fell into the hands of the battle groups of the 9. and 10. SS-PD.

Two days later the German front had firmed up between Hertogenbusch (right) and Maastricht (left). The 9. SS-PD. was on the right wing of the new defense front. The division remained there until the beginning of October, when it was transferred to Westphalia to rest and reequip with new Panther tanks. At the beginning of December the division drove through heavy snow over ice-covered roads into the Eifel in preparation for the Ardennes offensive.

When the offensive began the division was in the second line. When the I. SS-Pz.K.'s attack bogged down in the initial phase of the offensive, the units of the 9. SS-PD. were ordered forward and committed on the left wing of the *6. SS-Pz.Armee*. The German advance got little farther, however, and on Christmas Eve all further attacks were called off.

The order for the entire *6. SS-Pz.Armee* to transfer to Hungary resulted in the 9. SS-PD. moving to the area north of Lake Balaton. After the attack toward the Danube and the relief of Budapest failed, at the beginning of March the units withdrew to Austria. In the final weeks of the war the division was attached to the XXII. AK. of the *2. Pz.Armee*.

Commanders:

Brig.F. Bittrich; Brig.F. von Treuenfeld; Ob.F. Stadler; Ob.F. Bock.

10. SS-Panzerdivision "Frundsberg"

Establishment of the 10. SS-Pz.Gren.D. was scheduled to begin on 15 February 1943. The new division was created in barracks and troop training grounds in the southwest of France. Until April the units remained under the direct command of *Heeresgruppenkommando "D"* in the reserve role. The division received the honorary title *"Karl der Große"*.

After completion of training and the first regiment-size exercises the division was placed under the command of the AOK 1. Now a combat division, it remained in southern France as an occupation unit and in September 1943 was moved to Marseille.

Hitler did not agree with the division's name and on 3 October 1943 ordered it changed to *10. SS-Pz.Gren.D. "Frundsberg"*. One month later the division was transferred to northern France into the area of the *15. Armee* and remained there until April. In the final weeks the division was reorganized as a panzer division and was renamed 10. SS-PD.

Then it was transferred to the headquarters of the II. SS-Pz.K. (*1. Pz.Armee*) in the southern sector of the Eastern Front. The division fought in the Ternopol area in April 1944 and in May and June took part in the *Armee*'s retreat to the Lvov area.

At the end of June the division was rushed to Normandy. In the final days of June the first units to arrive – without heavy weapons and vehicles – were thrown into the Battle of Caen already under way. The remaining elements detrained in Paris and Metz and subsequently had to march on foot, under continuous air attack, to join the division's first battle groups.

When the German *7. Armee* was encircled at Falaise, the 10. SS-PD. was positioned on the *Armee*'s left wing in the Domfront area until mid-August. During the breakout from the pocket at the end of August the SS soldiers often provided the rear guard for elements of the army. The division lost most of its tanks and guns and was transferred with the 9. SS-PD. to the Netherlands to rest and reequip.

In mid-September Anglo-Canadian forces launched a large-scale airborne operation aimed at seizing the Rhine bridges. The units of the division were quickly committed to counterattacks against the enemy-held bridges in and around Arnhem. On 22 September the division's battle groups stormed the last bridges in Arnhem and Oosterbeck held by British paratroopers, ending the Battle of Arnhem.

The subsequent retreat brought the 10. SS-PD. back into the Düren area as part of the LXXXVIII. AK. of the *15. Armee*. There was a brief respite in November, then the division was moved to the Eifel Region. The 10. SS-PD. did not take part in the Ardennes offensive. Instead it was transferred to Alsace to take part in the planned "Operation North Wind."

While still en route the division was stopped and turned around for priority transport to the newly-formed *11. Armee* in Pomerania. One month later the 10. SS-PD. was on the Oder front. There the 10. SS-PD., the 4. SS-Pol.D., 28. SS-D. and the *SS-Pz.Verband "Holstein"* made up the XXXIX. Pz.K. on the right wing of the *11. Armee*. The corps' mission was to guard the narrows between two lakes. A planned attack to improve the main line of resistance was unsuccessful.

After the failure of the limited German offensive in the Arnswalde area the division was transferred again, this time south to the *4. Pz.Armee* in Lausitz. The division was under the direct control of the OKH as its operational reserve and following the breakthrough by Russian armored forces between Cottbus and Spremberg in mid-April 1945 it was moved forward to Spremberg.

This began the division's final chapter. The SS men held out in the ruins of Spremberg until 22 April 1945. Then they were simply overrun by Soviet tanks. A number of groups were able to break out to the west and southwest. Several of these reached Dresden, others the Ore Mountains and still others marched in the direction of Eger and Karlsbad.

Commanders:

Brig.F. Debes, Gruf. von Treuenfeld; Brig.F. Harmel.

11. SS-Freiwilligen-Panzergrenadierdivision "Nordland"

In the summer of 1943 the *SS-Führungshauptamt* began combining the scattered SS volunteer units into larger formations. The *Legion "Norwegen"* and *Freikorps "Danmark"*, which had been operating as independent units within the army since the start of the eastern campaign, were pulled out of the front in spring 1943 and transferred to the Grafenwöhr troop training grounds.

The first volunteer division of the *Waffen-SS* was created there and was given the name *"Nordland"*. The former Norwegian Volunteer Regiment 1 became the SS-Pz.Gren.R. 23 *"Norge"*, while the Danish Volunteer Regiment 1 became the SS-Pz.Gren.R. 23 *"Danmark"*. Also created was the SS-Pz.Abt. 11 *"Hermann von Salza"* and all the necessary division units with the number 11.

After completing the necessary training, in September the division was transferred to the area of the *Wehrmacht* commander of Croatia. Then, in December, the division was transferred to the northern sector of the Eastern Front where volunteers from northern Europe had been fighting since 1941.

The division was sent directly to the front in the Oranienbaum bridgehead and took over the positions of the 10. Luftw.Feld.D. On 14 January 1944 the Soviets launched a major offensive there with the goal of freeing Leningrad. After a massive bombardment which saw 100,000 shells fired in a few hours, seven rifle and two tank divisions attacked the positions of four German divisions.

The men of the *"Nordland"* were able to defend their main line of resistance on the first day, however on the second day of the battle they were forced to give ground as Soviet tanks were already in their rear. As part of the III. SS-Pz.K. the division withdrew to the Narva by the end of January and the SS units dug in around Kingisepp. This was followed by a withdrawal into the so-called "Narva Position".

The units of the division saw action in all parts of the Narva bridgehead, depending on where the main weight of the Soviet offensive lay. Not until mid-September, when the order was received to evacuate Estonia, did the division's regiments and battalions withdraw.

In mid-September 1944 the units reached the Dvina near Riga and from there entered Courland. The river was crossed on 3 October 1944 and a few days later the division was inserted into the front northwest of Doblen to defend against a Soviet offensive. The division took part in the First Battle of Courland and maintained its positions. The Second Battle of Courland, which began at the end of October 1944, saw the 11. SS-Pz.Gren.D. in action between Skuodas and Vainode, with the division several times taking the brunt of the Soviet offensive. It also took part in the Third Battle of Courland, by which time it was 1945.

At the end of January 1945 the Red Army launched its offensives against the German Reich, after an earlier offensive had brought its tanks to Silesia. The III. SS-Pz.K. (11. and 23. SS-Pz.Gren.D.) was withdrawn from the front of *Heeresgruppe Kurland* (Army Group Courland) and transferred to the front in Pomerania.

In Pomerania the division fought as part of the *3. Pz.Armee*. Soviet forces breached the Oder front and advanced into Pomerania, Brandenburg and Saxony. At the end of April 1945 the remnants of the division fought their way back into the greater Berlin area and together with attached volunteers of the *Division "Charlemagne"* participated in the final battles in the Reich capital. The battle groups of the 11. SS-Pz.Gren.D. saw action on the Unter den Linden, at the Halle Gate, at the Landwehr Canal, at Anhalt Station and the area of the Reich Chancellery. On 2 May 1945 the military commander of Berlin proposed surrender and the 11. SS-Pz.Gren.D. ceased to exist.

Commanders:
Brig.F. von Scholz; Brig.F. Ziegler; Brig.F. Krukenberg.

12. SS-Panzerdivision "Hitlerjugend"

The division was formed as a panzer-grenadier division in Turnhout, Belgium in July 1943 (plans dated 10 April 1943). Later, on 21 October, it was converted into a panzer division on Hitler's express order. The units were made up of Hitler Youth leader schools and cadre personnel from the *"Leibstandarte"*. Accordingly, on 22 October the division was given the name *"Hitlerjugend"* (Hitler Youth). All of the division's units were assigned the number 12 except for the two panzer-grenadier regiments which were numbered 25 and 26. The Pz.Jäg.Abt. 1 of the *"Leibstandarte"* was attached to the division but retained its number 1.

In April 1944 the new division left its quarters and moved to the Evreux area at the disposal of the OKW. On 6 June 1944 the division was placed on alert following the Allied landings in Normandy.

At 0700 hours orders were received for the division to proceed immediately to Rouen. The division crossed the Orne and in the morning advanced down the Villers-Bocage road to Caen, which had been evacuated by the *Luftwaffe* units stationed there. The Battle of Caen began three days later. A report stated:

> "The 12. SS-PD. *"Hitlerjugend"* saw its first action west of the city. It was to become the most feared division in those days, even though its soldiers were no older than eighteen on average. In its first live action on 7 June the division destroyed 28 Canadian tanks and took 150 prisoners."

Heavy fighting developed between Tilly and Caen which lasted for days and claimed heavy casualties on both sides. On 28 June the division held its ground against a fresh enemy offensive against Caen. The division's battle groups held out for eight days in the ruined city and at Carpiquet airfield.

The defense front broke on the morning of 8 July 1944. Two British and one Canadian divisions attacked the positions of the 12. SS-PD, which by this time had been reduced to several battle groups. The I./SS-Pz.Gren.R. 25 fought to the last man. Only 25 men of the same regiment's III. Btl. survived and none of its officers.

Feldmarschall von Rundstedt, Commander-in-Chief West, said in a later order of the day:

> "These soldiers possess the enthusiasm of the young regiments of Langemarck, however their training is vastly superior… It is a shame that these trusting youths are being sacrificed in a hopeless situation."

The subsequent retreat brought the division's battle groups to Falaise, where they were surrounded with the *7. Armee*. The 12. SS-PD. defended Falaise to the end. By the last day of the battle for the city only 60 SS men and two tanks were left. By about midnight on 17 August not one was left alive. After the breakout from the pocket the division was reduced to just ten tanks and six guns.

The division was withdrawn from the front and sent to the Eifel and later to the area around Bremen. The division was rested there until the end of November and its units received replacements, new weapons and vehicles. Then the division returned to the front.

When the Ardennes offensive began on 16 December 1944 the 12. SS-PD. was again in the forefront and in the eight days that followed it advanced through the snow-covered forests and over icy roads. The American 1st and 30th Infantry Divisions put up stubborn resistance. The division reached the area due south of Malmédy but could go no farther. Its strength was spent. The division's objective – the bridges over the Meuse near Liège – could not be reached. The retreat which began in January 1945 took the division via Bastogne to the border of the Reich. The 12. SS-PD. was taken out of the front and was reequipped in the Cologne area. Then, at the beginning of February, it was transferred to the *6. SS-Pz.Armee* in Hungary.

There the division participated in the confused fighting at Köbölkut and in the Gran bridgehead. Then it was moved to Ödin-Puszta and the Sio Canal. In March the division participated in the defense of Stuhlweissenburg, after which it withdrew via the Margarethe Position to the Raab. From there the division proceeded to Ödenburg, where it was temporarily encircled. In April 1945 it conducted a fighting withdrawal to the Vienna Woods and took part in heavy fighting in the Altenmarkt area.

At the beginning of May 1945 the unit reached the Enns, several battle groups made it as far as Linz. Then the war ended for the 12. SS-PD. *"Hitlerjugend"*.

Commanders:
Gruf. Witt; Brig.F. Meyer; Ob.F. Kraemer; Brig.F. Kraas.

13. Waffen-SS-Gebirgsdivision "Handschar" (Kroatische Nr. 1)

On 1 March 1943 Hitler ordered the formation of a division of volunteers from Croatia. Since 1941, following the establishment of the German administration in Croatia and Bosnia, many young men had volunteered for service in the *Wehrmacht*. The formation of a division-size unit was delayed, as the expected influx of Muslims did not materialize. On 2 July 1943 the existing volunteer units were combined to form the *Kroatische SS-Freiwilligen-Geb.D.* (Croatian SS Volunteer Mountain Division). Three months later on 22 October, the division was renamed the *13. SS-Freiwillige Bosnisch-Herzogowinische Geb.D.* (13th SS Volunteer Bosnian-Herzogovinan Mountain Division).

While in southern France completing the establishment process, the division received its ultimate title *"Handschar"* with the suffix *"kroatische Nr. 1"* (Croatian No. 1).

At the end of December 1943 and beginning of January 1944 the division was transported to Croatia. In April 1944, under the command of the *2. Pz.Armee*, it began operations against the partisans which were becoming increasingly active.

At the beginning of June 1944 additional partisan battalions began entering western Serbia from eastern Serbia and the Sandschak. The division's battle groups fought pitched battles for strongpoints northeast of the Lim and on the west bank of the Drina.

The months of fighting in the rocky mountain terrain led to heavy losses in men and materiel. When the Red Army finally entered Bulgaria and Romania, the division was deployed on the south bank of the Save. The volunteers fought with determination, however when the retreat into their homeland began in October (attached to the IX. SS-AK.) the ability of the division units to resist dropped noticeably.

Under the command of the LXVIII. AK. of the *2. Pz.Armee*, the division was forced to withdraw to Hungary. For the next three months the remnants of the division fought the advancing motorized units of the Red Army. By the beginning of May several battle groups managed to reach Carinthia, where they were taken prisoner by the British following the German surrender.

Commanders:
Gruf. von Sauberzweig; Ob.F. Hampel.

14. SS-Waffen-Grenadierdivision (Galizische Nr. 1)

The order to form a grenadier division from volunteers from the Ukraine and Galicia was issued on 28 April 1943. It was not until November 1943, however, that the division was established at the Heidelager troop training grounds in Debica near Krakow. It consisted of 13,500 officers, non-commissioned officers and enlisted men. The latter were for the most part ethnic Germans and Ukrainians from the Lvov area. Effective 22 October 1943 the division was named the *14. Waffen-Gren.D. (Galizische Nr. 1)*. The division's three regiments were numbered 29 to 31.

The establishment process was completed by summer 1944 in the rear area of Army Group South, especially in the East Ukraine. Following the start of the Soviet offensive in June 1944 the division was attached to the *1. Pz.Armee*. As a result it was trapped in the Brody pocket with the entire XIII. AK.

The soldiers of the division fought bravely, especially on the road between Pidhirtsi and Zolochiv. The men of the Gren.R. 30 distinguished themselves through their outstanding bravery. In the end, however, only 3,000 members of the division escaped from the pocket.

The SS units which escaped the pocket were sent back to the Neuhammer troop training grounds. A capable new combat unit was created there and on 28 September 1944 it was transferred to Slovakia as an occupation unit. After the Red Army's advance at the turn of the year, the division was moved to Sillein, where it was renamed the *14. Waffen-Gren.D. der SS (Ukrainische Nr. 1)*.

A few days later the division moved into the Pskov area and as part of the I. Kav.K. of the *2. Pz.Armee* fought its way back to Styria. There the battle groups were captured by American forces near Tamsweg and Judenburg.

Commander:
Gruf. Freitag.

15. SS-Waffen-Grenadierdivision (Lettische Nr. 1)

The division was created from the *"Lettischen Freiwilligen Legion"* (Latvian Volunteer Legion) which had been formed in Latvia on 10 February 1943. The legion consisted of the 1st-3rd volunteer regiments. When the units of the *Waffen-SS* were reorganized and numbered, the unit was expanded into a "new style" grenadier division. The grenadier regiments were renumbered 32 to 34, while the division itself was designated the *15. SS Waffen Grenadierdivision* with the suffix *"Lettische Nr. 1"* (Latvian No. 1).

In November-December 1943 the division was transported to the northern sector of the Eastern Front and assigned to the XXXXIII. AK. of the *16. Armee* on the Volkhov. When Soviet forces succeeded in forcing back the *Armee*'s right wing near Nevel, raising fears that Army Groups North and Center might be split apart, the division was moved into the area west of Velikiye Luki on the *Armee*'s right wing.

The history of the armed units of the SS began with the formation of a unit intended solely for the protection of Reichskanzler Hitler. Its members received military training with the support of the Reichswehr. The term Waffen-SS came into being in 1940; prior to this the organization's military units were called the "SS-Verfügungstruppe".

The first Sturm of the "Leibstandarte" under Untersturmführer Witt (KIA near Caen on 12/06/1944 as commander of the 12. SS-PD.) on the grounds of the barracks installation in Berlin-Lichterfelde.

During the prewar period the units of the SS-Verfügungstruppe were used only as honor formations and security units. Above: Honor formation of the SS-Standarte "Germania" during the Duke of Windsor's visit to the Ordensburg Crössinsee. Below: Hitler inspects the "Leibstandarte". Beside him is SS-Obergruppenführer Dietrich.

The entry of the "Leibstandarte" into Saarbrücken on 1 March 1935 (above) and the Sudetenland in October 1938 (below). Note the field gray uniforms in the bottom photo.

During the Polish campaign the units of the SS-Verfügungstruppe saw action with various army divisions. Here SS companies move into a Polish village behind an armored scout car.

In the western campaign all units of the SS-Verfügungstruppe and the Polizei-Division saw action in the front lines. The photo shows members of the Pol.D. in police uniforms waving to the enemy across the Rhine. This was a feature of the "Sitzkrieg" (Phony War) before the German offensive.

An SS assault team crosses a river in central France, May 1940 (photo above).

"Leibstandarte" victory parade in Berlin following the western campaign.

Motorcycle troops of the "Leibstandarte" during the rapid advance to Mount Olympus. The "Leibstandarte" and "Das Reich" Divisions took part in the campaign in the Balkans, with the "Leibstandarte" fighting the Greek and British armies and "Das Reich" the Yugoslavian army.

On 20 April 1941 SS-Obergruppenführer Dietrich conducted surrender negotiations with representatives of the Greek Commander-in-Chief.

When the eastern campaign began on 22 June 1941 every division of the Waffen-SS was committed in the front lines from the northern sector of the Eastern Front to the southern sector. An assault team from the advance detachment of "Das Reich" crossing the border early on the morning of the first day of the campaign.

Motorized units of the "Das Reich" pursue the enemy in the central sector of the Eastern Front.

An infantry gun of the "Totenkopf" Division engages Russian tanks in the forest near Staraya Russa.

Not only did the early onset of winter in 1941 halt the German offensive everywhere, it was also followed by the Soviet counteroffensive which drove back the German army on every front. The photo above illustrates the difficulty of transporting supplies in the southern sector of the Eastern Front. The photo below was taken on the battlefield at Rzhev.

The SS-PD. "Wiking" helped spearhead the German summer offensive toward the Caucasus in 1942. By the end of the offensive it was far ahead of the other units of the 1. Pz.Armee. Above: SS-Obergruppenführer Steiner, the division commander, in a slit trench at the edge of a sunflower field somewhere in the steppe. Below: A 37-mm anti-tank gun in firing position at an elevation of 2,000 meters in the Near-Caucasus.

Contrary to Hitler's orders, on 7 March 1943 the newly-formed SS-Pz.K. seized the initiative, drove into Kharkov with its three divisions and stabilized the front in the Ukraine (above). In the summer of 1943 the German command launched "Operation Zitadelle" in an attempt to reverse the tide on the Eastern Front. Panzer divisions of the Waffen-SS (below) played a major part in the offensive.

The Waffen-SS volunteer regiments made up of men from northern and eastern Europe were deployed almost exclusively in the northern sector of the Eastern Front, the most desolate of all. These regiments were later expanded into brigades and divisions. The photo shows a corduroy road in the Volkhov sector.

The III. SS-Pz.K. defended a sector in the Narva bridgehead and provided the Red Army with its most serious opposition in the northern sector.

In the last two years of the war volunteers from all parts of Europe were taken into the Waffen-SS where they served more or less bravely in action against the Red Army or partisans, especially in southeastern Europe.

Above: Soldiers of the SS-Geb.D. "Prinz Eugen" fought against Tito's partisans. Below: The Mufti of Jerusalem inspects an SS unit of Moslems.

The SS divisions' next major commitment followed the start of the Allied invasion of Normandy. The 12. SS-PD., much feared by the enemy, played a central role in the Battle of Caen (above). The Commander-in-Chief of the German 7. Armee in Normandy was SS-Oberstgruppenführer Hausser (center), the highest ranking officer of the Waffen-SS. To his left is General der Fallschirmtruppe Meindl.

The last major operation by the SS panzer divisions took place during the Ardennes offensive in December 1944. Above: Assault guns of the "Leibstandarte" attacking toward Malmedy. Below: Men of the Waffen-SS remained in combat until the end of the war. Here a knocked-out armored troop carrier in the wrecked streets of Berlin, May 1945.

The final resting place for thousands of soldiers of the Waffen-SS was a modest grave somewhere in Europe—here at the edge of a nameless village in the vast Ukraine.

The division remained on the right wing of the *16. Armee*, from January 1944 in the sector of the VI. SS-AK. Following the start of the major Soviet offensive on the Leningrad and Volkhov front in January 1944, the *16. Armee* was forced to join Army Group North's withdrawal. By March 1944 the division reached the area around Pskov and in April was pushed back to Nevel, where it held position. There, on 22 June 1944, the Red Army launched a major offensive which split Army Groups North and Center.

In mid-July 1944 the 2nd Baltic Front attacked the right wing of the *16. Armee*, shattering the German front near Novosokolniki. The 15. SS-Gren.D. was caught up in the maelstrom of the retreat and by the end of July its had ceased to exist as an effective division.

The army group sent the division from the main line of resistance to Livland and from there to West Prussia. There the 15. Waffen-Gren.D. was reformed in the period from October 1944 to January 1945. Before the reestablishment process could be completed, however, Soviet tanks had already driven to the borders of the Reich. In February 1945 the division was briefly attached to the XVIII. AK. of the *2. Armee* in West Prussia, then it was transferred to East Pomerania and in March was attached to *General* Tettau's Korpsgruppe. The latter was supposed to defend the Pomeranian border area in the Schneidemühl area as part of the *3. Pz.Armee*. The 15. SS-Gren.D. was split up into battle groups, some of which made it back to the Pomeranian coastline, while others even got as far as Courland, which was home to the Latvian volunteers.

Commanders:
Ob.F. Heilmann; Stand.F. Count von Pückler; Gruf. Pfeffer-Wildenbruch; Brig.F. Schuldt; Ob.F. Obwurzer.

16. SS-Panzergrenadierdivision "Reichsführer-SS"

This division was formed from the former *SS-Sturmbrigade Reichsführer-SS* in Laibach on 3 October 1943. Its predecessor was the guard battalion of the *Reichsführer-SS*, which on 14 February 1943 had been sent to Corsica as an occupation unit. These were the first German soldiers to set foot on the French island. In the months that followed, the battalion was expanded to brigade size and on the day of Italy's surrender was near Sarténe in the south of the island. At the same time, at Bonifacio the anti-tank battalion covered the crossing of the 90. Pz.Gren.D. from Sardinia to Corsica. When the Italian forces occupying the island refused to surrender, the brigade stormed the capital, Bastia.

The German troops were subsequently ordered to depart Corsica and leave the island to the French, however. At the end of September the *Sturmbrigade* was transferred to northern Italy and on 3 October 1943 it was expanded into a panzer-grenadier division at Laibach. The division's two regiments were numbered 35 and 36.

In October 1943 Allied forces landed at Anzio-Nettuno. The new division immediately rushed battalions to the eastern wing of the bridgehead. At the turn of the year the bulk of the 16. SS-Pz.Gren.D. was transferred to Baden near Vienna for training, while several battle groups were sent to Debreczen.

In May 1944 the entire division was transported to northern Italy and in July it was attached to the newly-formed *Armeegruppe von Zangen*. The division saw its first action in the Grosseto sector near Livorno. Its mission was to repulse a possible landing of Allied troops from Elba.

The division saw its first action in the new sector on 9 July 1944 near the town of Volterra southeast of Livorno. The division lost 49 officers in the fighting there. The subsequent retreat saw the division cross the Arno into the Carrara area on 19 July. There was further defensive fighting there as part of the XIV. Pz.K. of the *14. Armee*. In autumn 1944 the division was in the sector of the I. Fallsch.Jäg.K. between Bologna and Florence. In heavy fighting south of Florence the division lost 823 officers, non-commissioned officers and enlisted men killed or wounded.

Before leaving Italy the division was attached to the LXXXIII. AK. of the *10. Armee*. Then in January 1945 the division was relieved and transferred to Hungary. There the regiments fought the onrushing armored divisions of the Red Army in the area of Nagykanisza. In the final months of the war the division was under the command of Headquarters, I. Kav.K. of the *2. Pz.Armee*. In April 1945 it withdrew into Styria with the corps and when the war ended was captured there.

Commanders:

Ob.Sturmb.F. Gesele; Gruf. Simon; Ob.F. Baum; Ob.F. Deisenhofer.

17. SS-Panzergrenadierdivision "Götz von Berlichingen"

In late autumn 1943 the independent SS-Pz.Gren.Brigaden 49 and 51 were pulled out of their positions and transported to southwestern France. There, together with units sent from the Reich, including the army's 10. PD., they formed the new 17. SS-Pz.Gren.D. which was named *"Götz von Berlichingen"*.

After a brief training period, in December 1943 the division was sent to the V. SS-Geb.K., part of the *2. Pz.Armee*, to take part in the struggle against Tito's partisans in the Balkans. However, after just a few weeks the division returned to southern France as OKW reserve.

When the Allies landed in Normandy on 6 June 1944, the division was placed on alert and in the days that followed was transferred to the St. Lô area. Two days later advance units arrived on the battlefield near Littry. The division was moved into the front south of Carentan and attacked north against three American divisions. Later it was moved into the front south of St. Lô, bolstering the German defense front. By the time the city fell to the Americans on 20 July the division was down to 1,000 troops in the front lines.

The unit had been bled white and a few days later was withdrawn from the front to rest and reequip in the Champagne and the Paris area. The division was subsequently sent to the *1. Armee* and in September 1944 returned to action in the Metz area.

By that time the American 3rd Army had launched an offensive against the Moselle sector. The 17. SS-Pz.Gren.D. occupied the front in Metz and just to the north of the city. The front there held until mid-November, when the Americans launched another offensive. This time the division – minus the Pz.Gren.R. 38 which was destroyed in Metz – had to pull back to the borders of the Reich.

"Operation Northern Light", the last German offensive in the west, began on 1 January 1945 in northern Lorraine. The 17. SS-Pz.Gren.D. was in the front line and on New Year's Day it attacked through Wölfingen and Binningen toward Achern on the right wing of the *1. Armee*. On the second day, however, the attack became bogged down in the old Maginot Line.

The subsequent retreat took the division through Baden, northern Württemberg and Bavaria. Battle groups of the division escaped American encirclement and reached Bavaria, arriving in the Munich area on 30

April 1945. The remnants of the division were in Upper Bavaria when the war ended.

Commanders:

Brig.F. Ostendorff; Ob.F. Baum; Stand.F. Binge; Ob.F. Deisenhofer; Stand.F. Lingner; Ob.F. Klingenberg; Stand.F. Bachmann.

Divisions Formed in 1944-1945

In terms of organization, equipment and personnel strengths, the divisions of the *Waffen-SS* formed between 1939 and 1943 were roughly equivalent to the divisions of the army. As the war progressed, however, both the army and *Waffen-SS* began to run short of experienced personnel as well as weapons and materiel. Consequently, large units formed in 1944 and 1945 could not match the strengths and equipment of the earlier divisions. This is not to say, however, that the officers, non-commissioned officers and enlisted men of the divisions formed in the later war years did not possess the same spirit and bravery as those of the older units.

The lives of the later divisions rarely exceeded one year, indeed some were in existence for just months or even weeks. Nevertheless, each of these divisions is described below. After all, the men of the 18th to 38th Divisions of the *Waffen-SS* "were soldiers, just like the others."

18. SS-Freiwilligen-Panzergrenadierdivision "Horst Wessel"

Formed on 25 January 1944 in the Agram-Cilli area from the former 1. SS-Gren.Br. with Rgter. 39 and 40. Attached to the *Wehrmacht* commander in Hungary as a reserve unit from July to October 1944. From November 1944 in action in the southern sector of the Eastern Front with the *6. Armee*, then to the *8. Armee* in Slovakia. February 1945 to the *1. Pz.Armee* in Upper Silesia and from April with the *17. Armee* in the final battle in Silesia. Some groups escaped encirclement near Neustadt and with three infantry divisions were the last German units to cross the passes to Bohemia.

Commanders:

Ob.F. Trabandt; Stand.F. Bochmann; Stand.F. Petersen.

19. SS-Waffen-Grenadierdivision (Lettische Nr. 2)

This was one of the volunteer units from the Baltic States and was created in Latvia as the *2. lettische SS-Brig.* (2nd Latvian SS Brigade) shortly after the beginning of the eastern campaign. In the months that followed, it saw action in the Leningrad area, later at the Volkhov, and in other sectors held by Army Group North. At the end of 1943 it was pulled out of the line and on 17 January 1944 was established as the 19. SS-Gren.D.

After a brief period of training it returned to the front with its three regiments and other division units. The troops fought at Pskov as part of the 16. AR, later in Livland and from October 1944 in Courland. The division participated in all six Battles of Courland and when the end came was located southwest of Tukkum.

Commanders:

Brig.F. Schuldt; Gruf. Streckenbach.

20. SS-Waffen-Grenadierdivision (Estnische Nr. 1)

The division was established on 24 January 1944 in the rear area of Army Group North from Estonian volunteer units, some of which had existed since 1941. It was conventional infantry division which included the SS-Gren.Rgter. 45, 46 and 47.

The division first saw action in March 1944 with the III. SS-Pz.K. on the Narva front. After three months in combat the division was considered battle-weary and was reformed in Neuhammer in November 1944.

After the start of the Soviet offensive against Silesia in January 1945 the division was rushed to the area and joined the *17. Armee* in February. That same month it withdrew into the Striegau area and was encircled by Soviet units near Neustadt. Battle groups fought their way out of the encirclement, crossed the border into Bohemia and were captured by the Russians near Melnik, north of Prague.

Commander:

Brig.F. Augsberger.

21. SS-Waffen-Gebirgsdivision 'Skanderberg" (Albanische Nr. 1)

The division was formed in the Pac-Pristina-Prizren area in northern Albania from Albanian volunteers, the majority of them Moslems, and consisted of the SS-Geb.Jäg.Rgter 50 and 51. The establishment process could not be completed as the division was committed against partisans with the XXI. AK. in Albania in June 1944. The division remained in constant action against these units, from October in northern Yugoslavia and in December in Croatia under the V. SS-Geb.K. The division was disbanded in November 1944 and the remaining units were incorporated into the 7. SS-Geb.D.

Commander:
Brig.F. Schmidhuber

22. SS-Freiwilligen Kavalleriedivision "Maria Theresia"

Formed from Hungarian-German volunteers in Hungary on 29 April 1944, the division initially consisted of two cavalry regiments (a third regiment from the 8. SS-Kav.D. was added later) and the usual division units. After completion of basic training in August 1944, the division was transferred into Army Group South Ukraine's rear area as a reserve unit. In October it went into action with the German *6. Armee* and fought against the Russian forces which had broken into Hungary from the east and south. The division took part in the fierce defensive fighting there and on Christmas Eve 1944 was encircled in the Hungarian capital together with other units of the army. When the hopeless struggle for the city ended on 12 February 1945 (all German relief attempts failed), a few of the division's soldiers succeeded in breaking out to the west.

Commander:
Brig.F. Zehender

23. SS-Waffen-Gebirgsdivision "Kama" (Kroatische Nr. 2)

The division was formed on 10 June 1944 in Croatia in the area Save-Bosna-Spreca-Drina. The unit, which was made up of Croats, consisted of the SS-Geb.Jäg.Rgter. 55 and 56 and the necessary division units. The division saw no action as it was disbanded on 24 September 1944. The Moslem soldiers were assigned to the 13. SS-Waffen-Geb.D. *"Handschar"* and the last cadres were disbanded on 31 December 1944.

23. SS-Freiwilligen-Panzergrenadierdivision "Nederland"

The second division to be assigned the number 23, this division was created from the *SS-Freiwilligen-Legion "Nederland"*, which had been in action since June 1943. Some of the units officers and men had been in combat in the northern sector of the Eastern Front since 1941. After establishment, in November the division moved to the Balkans as a reserve unit. In January 1944 it returned to the Eastern Front.

In January 1944 the first of the division's battle groups arrived at the front and were thrown into the battle raging in the Narva bridgehead. The division was deployed on the left wing of the bridgehead and, in spite of the enemy's superiority, defended its positions until spring 1944.

The Soviets launched their major offensive against the German Narva position on 24 July 1944. Two days later the division was forced to abandon the city and withdrew to the "Tannenberg Position". The *"Nederland"* Division fought fiercely for every meter of ground. By mid-August the front had reached Lake Peipus and the division assumed responsibility for defending the south wing, repulsing Soviet attacks against Dorpat.

The retreat to Courland began in mid-September. The III. SS-Pz.K., to which the division had been attached since January 1944, was transferred to Pomerania. In mid-February 1945 the division fought between Neudamm, Küstrin and Arnswalde, later southeast of Stargard, near Köslin and at Stettin. By this time the division had been reduced to the strength of a battle group, which was able to fight its way through to the coast.

Commanders:
LtGen. (Dutch Army) Seyffarth; Brig.F. Wagner.

24. SS-Waffen-Gebirgsdivision "Karstjäger"

Formation of the division began on the Adriatic coast on 1 August 1944. The local peoples provided the personnel for the SS-Karstjäg.R. 59, an artillery battalion, a pioneer company and a signals company. Four months later the Karstjäg.R. 60 was formed, however the unit never reached division strength. It was employed in the anti-partisan role for a time and in 1945 was withdrawn to Istria. There it was employed behind the front and on 5 May 1945 was captured by the Allies.

Commanders:
Stand.F. Obwurzer; Ob.F. Heilmann.

25. SS-Waffen-Grenadierdivision "Hunjadi" (Ungarische Nr. 1)

Beginning on 2 November 1944 the Hungarian volunteers fighting in the ranks of the *Waffen-SS* were assembled at various troop training grounds in the Reich. They formed the new SS grenadier division which consisted of the Gren.Rgter. 61, 62 and 63, other division units and a ski battalion. All of these units were brought together at the Neuhammer troop training grounds by the beginning of 1945.

Following the start of the Soviet offensive against Silesia in January 1945, battle groups from the division were sent into action at the front. The division did not see action as a unit. Attached to other divisions, its battle groups withdrew over the mountains and through Bohemia and by the beginning of May 1945 reached the Nuremberg area.

Commanders:
Stand.F. Müller; Ob.F. Heilmann

26. SS-Panzerdivision

On 10 August 1944 the 49. SS-Pz.Gren.Brig., which had previously operated independently, was increased to division strength and became the 26. SS PD. In those weeks the division was attached to the German *7. Armee* in the Champagne area of France. On 8 September 1944 the division was disbanded.

26. SS-Waffen-Grenadierdivision "Gömbos" (Ungarische Nr. 2)

The division was formed in January 1945 from the 49. SS-Pz.Gren.Br. (see 26. SS-PD.). It was attached to the Commander of the Replacement Army in the Home War Zone and saw no front-line action.

Commander:
Stand.F. Gross.

27. SS-Panzerdivision

On 10 August 1944 the SS-Pz.Brig. 51, previously employed in the west, was renamed the 27. SS-PD. and as such was employed in the Champagne region for several days under the command of the *1. Armee*. But on 8 September 1944 it was ordered disbanded. Its personnel were assigned to the 17. SS-Pz.Gren.D.

27. SS-Freiwilligen Grenadierdivision "Langemarck"

The 6. SS-Sturm-Brig. *"Langemarck"* saw action with Army Group North in Courland. At the end of October 1944 it was pulled back into the Soltau area. There the brigade was expanded to form the 27. SS-Gren.D. with Rgter. 66, 67 and 68, Pz.Abt. 67 and all division units.

In January 1945 the new division went to the front in Pomerania and was attached to the III. SS-Pz.K., later to the *3. Panzer-Armee*. The division took part in the fighting in West Prussia and later on the Oder front. After taking heavy casualties it pulled back to Stettin and was captured there by the Russians on the day of the surrender.

Commanders:
Stand.F. Schellong; Gruf. Berger.

28. SS-Freiwilligen-Grenadierdivision "Wallonien"

The division, which was formed in the south Hanover area on 19 October 1944, consisted of the Gren.Rgter. 69 and 70 plus division units with the number 28. The majority of its soldiers had served with the SS-Brig. *"Wallonien"*, which had been deployed on the Eastern Front in northern Russia since 1941.

In December the division was redesignated a panzer-grenadier division and formation was completed by January 1945, after which it was assigned to the *11. Armee* in Pomerania.

The division fought on the *Armee*'s right wing under the command of the XXXIX. Pz.K. On 15 February 1945 the division took part in offensive operations in the Stargard area. The advance bogged down east of Lake Madü, however. The regiments were assigned to the III. SS-Pz.K., which successfully defended the front line between Stargard and the Oder. Stargard was abandoned on 4 March and the retreat to the Oder began.

At the end of March *"Wallonien"* saw action between Greifenhagen and Stettin. The battered division was subsequently withdrawn from the front and attached to Army Group Vistula as its reserve unit. The fighting withdrawals brought the SS units to Brandenburg, where the bulk of them were captured by the Soviets. Only a few SS companies managed to make their way to Denmark.

Commander:
Stand.F. Degrelle.

29. Waffen-Grenadierdivision "Rona" (Russische Nr. 1)

The division was supposed to be formed in Italy in August 1944 from the former Sturmbrig. *"Rona"*, whose members were all Russian volunteers. In October 1944, however, the unit was disbanded before formation was complete.

29. SS-Waffen-Grenadierdivision (Italienische Nr. 1)

The *SS-Gren.Brig. Italienische Nr. 1* was supposed to be expanded into a conventional grenadier division beginning in February 1945. Two grenadier regiments and a few division units were subsequently established, however the end of the war meant that the new division never saw action.

Commander:
Brig.F. Hansen.

30. SS-Waffen-Grenadierdivision (Weißruthenische Nr. 1)

The *Schutzmannschafts-Brig. Sieglinge* was made up of Russian, Ukrainian and White Russian volunteers and saw action in the police role in White Russia and later Poland. On 1 August 1944 it was expanded into a division with three grenadier regiments. In October 1944, with the Red Army approaching, the as yet incomplete division was transferred to Belfort.

After seeing action briefly in the *19. Armee*'s sector on the Upper Rhine, the unit's soldiers suddenly mutinied, shot their officers and went over to the enemy. What was left of the unit was immediately withdrawn from the front and transported to Grafenwöhr. There the division was disbanded in January 1945 and its members were transferred to the "Vlasov Army".

Commander:
Stand.F. Sieglinge

31. SS-Freiwilligen-Grenadierdivision "Böhmen-Mähren" (Bohemia-Moravia)

On 1 October 1944 the disbanded *23. SS-Waffen-Gren.D.* was assembled in the Fünfkirchen area of Hungary. The *SS-Führungshauptamt* used its personnel to create the new 31. SS-Gren.D. One-thousand Moslem Bosnians of the old division were transferred to the 13. SS-Geb.D. After the addition of men from South Tyrol the new division's formation was completed in Marburg/Drau.

In November the division was attached to the Hungarian 2nd Army and later to the *2. Pz.Armee*. It took part in the latter's fighting withdrawal into southern Styria. In April 1945 the division was transferred to Silesia and during the final weeks of the war it fought between Striegau and Strehlen under the XVII. AK. of the *17. Armee*.

Commander:
Brig.F. Lombard.

32. SS-Freiwilligen-Grenadierdivision "30. Januar"

This division, which consisted of two grenadier regiments and the usual division units, was created at the Jamlitz troop training grounds in the Lublin District on 30 January 1945. Its personnel came from various schools and recruiting depots of the *Waffen-SS*, including the Courland

Recruiting Depot, the SS Artillery Training Regiment in Prague, the SS Pioneer School in Hradischko and Battle Group Schill in Brünn.

Several weeks later the new unit went into action in the Frankfurt/Oder area as part of the V. SS-Geb.K. (*9. Armee*). The division was quickly thrown into the *Armee*'s fighting withdrawals to Brandenburg. It became involved in fierce actions in the Müllrose area and in the Halbe pocket southeast of Berlin. One battle group was forced back to Angermünde.

At the end of March 1945 the division reported a combat strength of just 2,846 men, having lost 407 killed, 1,447 wounded and 336 missing. Only a few of the SS men managed to make their way to the Elbe.

Commanders:
Ob.F. Mohnke; Stand.F. Mühlenkamp; Stand.F. Kempin.

33. SS-Waffen-Grenadierdivision "Charlemagne"

French citizens began volunteering for service in the German armed services following the western campaign in 1940. One French unit, Gren.R. 638, fought its way to Moscow in 1941. It was subsequently transferred to the *Waffen-SS* and was expanded into the *Brigade "Charlemagne"*. On 10 February 1945 the brigade formed the basis of the 33. Waffen-Gren.D. of the same name.

Before formation of the division was complete, it was sent to the *3. Pz.Armee*, which was engaged in fierce fighting in and around Pomerania. The new division was moved to east of Neustrelitz on 25 March 1945 and it subsequently pulled back to Köslin in fierce fighting. The unit was surrounded there. Some battle groups managed to fight their way through to Kolberg, Schwerin and Neustrelitz. One large group under the commanding officer reached Berlin, where it was used to defend the southeast part of the city against advancing Soviet armor. The French troops fought valiantly in the streets and houses, finally surrendering in the area of the Reich Chancellery. The last two soldiers to be awarded the Knight's Cross were French!

Commander:
Brig.F. Krukenberg.

34. *SS-Freiwilligen-Grenadierdivision "Landstorm Nederland"*

Brought back to the Netherlands from Army Group North in Courland, in February and March 1945 the *SS-Gren.Brig. "Nederland"* was expanded into a division with the SS-Gren.Rgter. 83 and 84. In April 1945 the division saw action on the right wing of the *15. Armee* in the Netherlands. As part of the XXX. AK., it fought its way back from Eindhoven to Arnhem, where it was captured by the Allies.

Commander:
Stand.F. Kohlrosser.

35. SS-Polizei-Grenadierdivision

This division was formed at the Oder front in February 1945 from the Pol.Rgter. zbV. 1 and 2 (Special Police Regiments 1 and 2), the Pol.Brig. Wirth and two *Volkssturm* battalions. Its regiments were assigned the numbers 89 to 91. The missing division units were taken from units of the former *Ordnungspolizei.* While still in formation the unit saw action on the Oder front. In mid-April the division had to abandon its positions near Lübben, after which it conducted a fighting withdrawal via Gross-Leuthen to Cottbus. The story of this division ended in the pocket which surrounded the *9. Armee* between Halbe and Baruth.

Commander:
Stand.F. Pipkorn.

36. *SS-Waffen-Grenadierdivision*

The division was created at the Oder front on 20 February 1945 and consisted of the former *SS-Brig. "Dirlewanger"* and elements of the army (2./Pz.Jäg.Abt. 681, Pi.Btl. 687, GR. 1244, etc.). The two newly-formed regiments and the division units saw action with the V. AK. (*4. Pz.Armee*) on the Lausitz area and were destroyed there.

Commander:
Ob.F. Dirlewanger.

37. SS-Freiwilligen-Kavalleriedivision

This third cavalry division was created from those elements of the 8. and 22. SS-Kav.D. which escaped from Budapest. They were joined by Hungarian and Romanian volunteer auxiliaries. Formation took place in the Pskov area. The division went into action immediately, fought between Znaim and Raab north of Vienna and was captured by the Americans near Zwettel.

Commander:
Stand.F. Gesele.

38. SS-Grenadierdivision "Nibelungen"

This last division of the *Waffen-SS* was not formed until April 1945, from officers and SS officer candidates of the *SS-Junkerschule* at Bad Tölz. The officer candidates formed the Rgter. 95, 96 and 97, the panzer battalion and the AR. 38 (artillery regiment). The unit was attached to the OKW as reserve in southern Germany but saw no action.

Commander:
Ob.F. Trabandt.

Appendices

Soldiers Awarded High-Ranking Decorations

Knight's Cross with Oak Leaves, Swords and Diamonds

Sequential Number	Rank/Name/Position	Date Awarded
12	SS-Gruf. Gille, C.O. 5. SS-PD.	17/04/1944
16	SS-Ob.Gruf. Dietrich, Com.Gen. I. SS-Pz.K.	06/08/1944

Knight's Cross with Oak Leaves and Swords

Sequential Number	Rank/Name/Position	Date Awarded
26	SS-Ob.Gruf. Dietrich, C.O. 1. SS-PD.	14/03/1943
39	SS-Ob.Sturmb.F. Dieckmann, C.O. Pz.Gren.R. "Westland"	10/10/1943
47	SS-Gruf. Gille, C.O. 5. SS-PD.	20/02/1944
56	SS-Ob.F. Schuldt, C.O. 2. lett. SS-Brig.	25/03/1944
65	SS-Brig.F. Prieß, C.O. 3. SS-PD.	25/04/1944
71	SS-Ob.Sturm.F. Wittmann, 2./SS-Pz.Abt. 501	22/06/1944
77	SS-Sturmb.F. Dorr, C.O. SS-Pz.Gren.R. 9	09/07/1944
83	SS-Gruf. Fegelein, C.O. 8. SS-Kav.D.	30/07/1944
85	SS-Ob.Gruf. Steiner, Com.Gen. III. SS-Pz.K.	10/08/1944
90	SS-Ob.Gruf. Hausser, CIC 7. Armee	26/08/1944
91	SS-Stand.F. Meyer, C.O. 12. SS-PD.	27/08/1944
94	SS-Brig.F. Wisch, C.O. 1. SS-PD.	30/08/1944
95	SS-Stand.F. Baum, C.O. 2. SS-PD.	02/09/1944
116	SS-Brig.F. Harmel, C.O. 10. SS-PD.	15/12/1944
119	SS-Ob.Sturmb.F. Peiper, C.O. SS-Pz.R. 1	11/01/1945
120	SS-Ob.Gruf. Krüger, Com.Gen. VI. SS-Pz.K.	11/01/1945
129	SS-Ob.F. Dörner, C.O. 4. SS-Pol.D.	01/02/1945
132	SS-Brig.F. Kumm, C.O. 7. SS-Geb.D.	17/03/1945
140	SS-Stand.F. Bochmann, C.O. 18. SS-Pz.Gren.D.	26/03/1945
150	SS-Ob.Sturmb.F. Weidinger, C.O. SS-Gren.R. 4	06/05/1945
151	SS-Ob.Sturmb.F. Wisliceny, C.O. SS-Pz.Gren.R. 3	06/05/1945
152	SS-Ob.F. Stadler, C.O. 9. SS-PD.	06/05/1945
153	SS-Ob.Gruf. Bittrich, Com.Gen. II. SS-Pz.K.	06/05/1945

In addition, members of the *Waffen-SS* were awarded the following high-ranking decorations:

Knight's Cross of the Iron Cross with Oak Leaves:	74
Knight's Cross of the Iron Cross:	465
(including 29 to foreign volunteers)	
Knight's Cross of the War Merit Cross with Swords	11

Comparison of *Waffen-SS* and German Army Officer Ranks

Waffen-SS	**Army**
SS-Oberstgruppenführer (Ob.Gruf.)	*Generaloberst*
SS-Obergruppenführer (O.Gruf.)	*General der Infanterie, Artillerie, etc.*
SS-Gruppenführer (Gruf.)	*Generalleutnant*
SS-Brigadeführer (Brig.F.)	*Generalmajor*
SS-Oberführer (Ob.F.)	*Oberst*
SS-Standartenführer (Stand.F.)	*Oberst*
SS-Obersturmbannführer (Ob.Sturmb.F.)	*Oberstleutnant*
SS-Sturmbannführer (Sturmb.F.)	*Major*
SS-Hauptsturmführer (Hpt.Sturmf.)	*Hauptmann, Rittmeister*
SS-Obersturmführer (Ob.Sturmf.)	*Oberleutnant*
SS-Sturmführer (Sturmf.)	*Leutnant*

Waffen-SS ranks always included the prefix SS-; this has been omitted in this chapter for reasons of space.

Abbreviations

AA	*Aufklärungsabteilung* (reconnaissance battalion)
Abt.	*Abteilung* (battalion)
AK.	*Armeekorps* (army corps)
AOK	*Armeeoberkommando* (army high command or headquarters)
AR.	*Artillerieregiment* (artillery regiment)
Art.	*Artillerie* (artillery)
Br.	*Brigade* (brigade)
Brig.	*Brigade* (brigade)
Btl(e)	*Bataillon(e)* (battalion, battalions)
Bttr.	*Batterie* (battery)
D.	*Division* (division)
d.	*der* (of)
Divn.	*Divisionen* (divisions)
Ers.	*Ersatz* (replacement)
Fallsch.	*Fallschirm* (parachute)
FEB	*Feldersatz-Btl.* (replacement training battalion)
Fl.	*Flieger* (airman)
Fl.D.	*Fliegerdivision* (air division)
Fl.K.	Fliegerkorps (air corps)
Geb.	*Gebirgs-* (mountain)
Geb.D.	*Gebirgsdivision* (mountain division)
Geb.Jäg.	*Gebirgsjäger* (mountain infantryman)
Geb.K	*Gebirgskorps* (mountain corps)
Geb.Tr.	*Gebirgstruppe* (mountain troops)
Gen.	*General* (general)
Gen.St.	*Generalstab* (general staff)
Glt.	*Generalleutnant* (lieutenant-general)
Gm.	*Generalmajor* (major-general)
Go.	*Generaloberst* (full general)
GR.	*Grenadier-Rgt.* (grenadier regiment)
Gruf.	*Gruppenführer* (*SS* lieutenant-general)
HJ	*Hitlerjugend* (Hitler Youth)
HKL	*Hauptkampflinie* (main line of resistance)

Hpt.Sturmf.	*Hauptsturmführer* (SS captain)
Hptm.	*Hauptmann* (captain)
Ia	*1. Gen.St.Offizier* (operations officer)
ID.	*Infanteriedivision* (infantry division)
Inf.	*Infanterie* (infantry)
IR.	*Infanterieregiment* (infantry regiment)
Jäg.	*Jäger* (private in mountain division)
Kav.	*Kavallerie* (cavalry)
Kdr.	*Kommandeur* (commander)
Kdr.Gen.	*Kommandierender General* (commanding general)
K.G.	*Kommandierender General* (*Waffen-SS* commanding general)
Kp(n)	*Kompanie(n)* (company, companies)
Kradschtz.	*Kradschützen* (motorcycle, as in motorcycle company, battalion, etc.)
lei.	*Leicht(e)* (light)
l.FH.	*leichte Feldhaubitze* (light field howitzer)
Lkw.	*Lastkraftwagen* (truck)
Ll.	*Luftlande-* (glider)
Ln.	*Luftnachrichten-* (*Luftwaffe* signal communication)
Luftw.	*Luftwaffe* (air force)
Luftw.Feld.D.	*Luftwaffen-Felddivision* (air force field division)
MG	*Maschinengewehr* (machine-gun)
Mj.	*Major* (major)
mot.	*motorisiert* (motorized)
NA.	*Nachrichtenabteilung* (signals battalion)
Nachr.	*Nachrichten-* (signals)
OB.	*Oberbefehlshaber* (Commander-in-Chief)
Ob.	*Oberst* (colonel)
Oblt.	*Oberleutnant* (lieutenant)
Oberstlt.	*Oberstleutnant* (lieutenant-colonel)
Ofw.	*Oberfeldwebel* (master sergeant)
Offz.	*Offizier* (officer)
OKH	*Oberkommando des Heeres* (army high command)
OKL	*Oberkommando der Luftwaffe* (air force high command)

OKW	*Oberkommando der Wehrmacht* (armed services high command)
OT.	*Organisation Todt* (Todt Organization)
PD.	*Panzerdivision* (armored division)
Pi.	*Pionier-* (combat engineer)
Pol.	*Polizei* (police)
Pz.	*Panzer* (armored, tank)
Pz.Abw.	*Panzerabwehr* (anti-tank)
Pz.Gren.D.	*Panzer-Grenadier-Division* (panzer-grenadier division)
Pz.Jäg.	*Panzerjäger* (anti-tank troops, tank destroyer)
Pz.K.	*Panzerkorps* (armored corps)
R.	*Regiment* (regiment)
RAD	*Reicharbeitsdienst* (Reich Labor Service)
Radf.	*Radfahr-* (bicycle)
Res.	*Reserve* (reserve)
Rgt.(er)	*Regiment(er)* (regiment, regiments)
San.	*Sanitäts-* (medical)
Scharf.	*Scharführer* (SS staff sergeant)
Stand.	*Standarte(n)* (SS regiment, regiments)
Sturmf.	*Sturmführer* (SS lieutenant)
Vers.	*Versorgungs-* (supply)
z.b.V.	*zur besonderen Verwendung* (special purpose)

NOTES

NOTES

NOTES

NOTES

NOTES

NOTES

NOTES

Also from the Publisher

Waffen-SS Commanders: The Army, Corps and Division Leaders of a Legend-Augsberger to Kreutz. Mark C. Yerger
The units of the Waffen-SS were some of the most successful and influential combat formations produced by any country in this century. Their abilities and accomplishments, in both defense and offense, remain legendary. Finally, the commanders of these elite units are examined here in detail. In this book, the first of a two volume set, forty-four biographies reveal the lives of the most senior Waffen-SS commanders.
Size: 8 1/2" x 11" • over 400 b/w photographs, maps, documents • 352 pp.
ISBN: 0-7643-0356-2 • hard cover • $59.95

Waffen-SS Commanders: The Army, Corps and Divisional Leaders of a Legend - Krüger to Zimmermann. Mark C. Yerger. In the second of a two volume set, sixty-one biographies reveal the lives of the most senior Waffen-SS commanders. Details are provided for education, as well as pre-Third Reich era service in military and civil posts, and includes promotions, assignments and decorations. The 1933-1945 era, the most detailed, reveals all their commands and related data similar to their earlier service. Also included is an addendum to Volume 1 that adds both text and photographic material uncovered during the concluding research.
Size: 8 1/2" x 11" • over 440 b/w photographs, 25 documents, glossary, appendix, index • 392 pp.
ISBN: 0-7643-0769-X • hard cover • $59.95

Himmler's Bosnian Division: The Waffen-SS Handschar Division 1943-1945. George Lepre. This is the story of the "Handschar," a Muslim combat formation created by the Germans to "restore order in Bosnia." What actually transpired was quite different.
Size: 6" x 9" • over 250 b/w photographs • 380 pp.
ISBN: 0-7643-0134-9 • hard cover • $39.95

The Camp Men: The SS Officers Who Ran the Nazi Concentration Camp System. French L. MacLean. Inside these pages you will meet over 960 infamous men – the officers of Nazi Germany's *Totenkopf* (Death's Head). These officers were not the bureaucrats who meticulously planned Adolf Hitler's Final Solution from behind a desk in Berlin, or those who quietly scheduled the trains that carried the victims to the camps. Quite the contrary; these men stood on the front-line of the Nazi war to exterminate the Jews. With well over one hundred photographs–a large portion previously unpublished— this is the largest collection of SS camp personnel photographs ever to appear in one work.
Size: 8 1/2" x 11" • over 140 b/w photographs, maps • 384 pp.
ISBN: 0-7643-0636-7 • hard cover • $59.95

The Field Men: The SS Officers Who Led the Einsatzkommandos - the Nazi Mobile Killing Units. French MacLean. Men lined up in four motorized columns immediately behind the German Army on June 22, 1941, as it prepared to launch Operation Barbarossa, an attack designed to win the war. The Field Men covers the entire gamut, from the organization of the units, to the SS officers who served in this scourge on the Eastern Front. Some 380 SS officers are described in full detail and extensively analyzed.
Size: 8 1/2" x 11" • over 175 b/w photos and maps • 232 pp.
ISBN: 0-7643-0754-1 • hard cover • $59.95

The Ghetto Men: The SS Destruction of the Jewish Warsaw Ghetto April-May 1943. French MacLean. The twenty-eight day siege of the Warsaw Ghetto was one of the most protracted large-scale urban battles in WWII. Only the fighting at Leningrad from 1941 1943, Stalingrad in 1942-1943, the Warsaw Rebellion of 1944, and Budapest in 1944-1945 lasted longer. This book presents every fact possible concerning the who, when, with what and how the SS troops razed the Warsaw Ghetto. The men and officers of the dreaded Security Service and Gestapo are here as well as the units, weapons and tactics, and a day-by-day analysis of the fighting.
Size: 8 1/2"x11" • over 110 b/w photos • 224 pp.
ISBN: 0-7643-1285-5 • hard cover • $59.95

SS Officers List (as of January 1942) SS-Standartfuhrer to SS-Oberstgruppenfuhrer - Assignments and Decorations of the Senior SS Officer Corps. The January 1942 "Dienstalterliste" lists the hundreds of officers in the SS from full Colonel to General. Detailing their post at that time during the war, this wartime publication, available now for the first time in a facsimile edition, also gives significant decorations they were awarded. Their ranks and most recent promotion dates are listed as well as State, Police and NSDAP posts. Illustrated with more than forty rare photos, this rare publication provides detailed information for historians and militaria collectors. Fully indexed.
Size: 8 1/2" x 11" • over 45 b/w photographs • 64 pages •
ISBN: 0-7643-0061-5 • soft cover • $19.95

In Perfect Formation: SS Ideology and the SS-Junkerschule-Tölz. Jay Hatheway. Includes extensive references to original source material on the underlying SS principles of blood, soil, and struggle as they were formalized in SS ideology. In support of his intricate linkages between ideology and its realized form, Hatheway has obtained over 100 previously unpublished photos of the SS officer training academy Tölz. More

than a series of buildings, the structure of the *Junkerschule* was itself a metaphor for the subset of Nazi ideology that was developed by Himmler, Darré and others to create a racially pure vanguard to lead Germany on its path toward Teutonic regeneration.
Size: 6" x 9" • over 100 b/w photographs • 192 pp.
ISBN: 0-7643-0753-3 • hard cover • $29.95

Himmler's Cavalry: The Equestrian SS, 1930-1945. Paul J. Wilson. Dressed in black uniforms emblazoned with the SS flash, Heinrich Himmler's horsemen carried the SS banner across the German countryside and to equestrian competitions throughout Germany and Europe. Wilson provides numerous anecdotes, statistics, and capsule biographies of Equestrian SS personnel. This carefully researched study is essential reading for scholars and lay people interested in the history of Nazi Germany, the SS, and the Holocaust.
Size: 6" x 9" • 17 b/w photographs • 224 pp.
ISBN: 0-7643-1112-3 • hard cover • $29.95

Twelve Years with Hitler: A History of 1.Kompanie Leibstandarte SS Adolf Hitler 1933-1945. by the 1.Kompanie LAH Veteran's Association, edited by Hans Quassowski. The original cadre of the later Waffen-SS was formed in March 1933 as the "SS Headquarters Guard Berlin." From the first 117 volunteers emerged more than fifty senior SS officers, all of whom received high decorations for bravery in the 38 Waffen-SS divisions that were formed later. Twenty former members of this company alone won the Knight's Cross. In the years of peace the 1st Company of this elite force took part in the most important functions of the Third Reich as an honor formation. This book is an impressive joint effort by surviving veterans in words and pictures.
Size: 9" x 12" • over 460 b/w photographs, 30 maps • 384 pp.
ISBN: 0-7643-0777-0 • hard cover • $69.95

Allgemeine-SS: The Commands, Units and Leaders of the General SS. Mark C. Yerger. The commands, units and leaders of the General SS are finally compiled into a single detailed reference book for both the historian and SS memorabilia collector. The biographical data for individuals alone adds vast detail to this fascinating topic. Along with more than 120 rare photos of SS senior ranking officers and seven maps, a detailed index allows referencing of individual commands or personalities.
Size: 8 1/2" x 11" • over 120 b/w photographs, maps • 256 pp.
ISBN: 0-7643-0145-4 • hard cover • $49.95

Heinrich Himmler: A Photographic Chronicle of Hitler's Reichsführer-SS. Martin Månsson. Many books have been written about Heinrich Himmler and the SS though, strangely, no photographic study has ever been assembled. Contained within this volume are over 380 photographs of Himmler, illustrating his entire career – from a soldier during the First World War period through the years when he was the second most powerful man in Europe. At the zenith of the Third Reich, all of Germany was under the controlling influence of Himmler and the SS. This photo chronicle records the entire Third Reich period in the life and death of Heinrich Himmler.
Size: 8 1/2"x11" • over 380 b/w photographs • 200 pp.
ISBN: 0-7643-1202-2 • hard cover • $59.95

Chronicle of the 7. Panzer-kompanie 1. SS-Panzer Division "Leibstandarte." Ralf Tiemann. This chronicle of the 7.Panzerkompanie follows the unit history of a "Leibstandarte" tank company from its creation in 1942 to the end of World War II. Compiled by former company commander and German Cross in Gold holder Ralf Tiemann, the detailed text relies on both official documentation and the personal recollections of numerous unit veterans. Exacting cartographic material allows the reader to follow all engagements in this first unit history of a "Leibstandarte" tank company available in English.
Size: 6" x 9" • over 60 b/w photos, over 30 maps, charts • 320 pp.
ISBN: 0-7643-0463-1 • hard cover • $29.95

Tank Combat in North Africa: The Opening Rounds Operations Sonnenblume, Brevity, Skorpion and Battleaxe. Thomas L. Jentz. The focus of this book is the tank battles fought during the opening phase of the war in North Africa. This book is compiled from the results of digging through original records for over thirty years in an attempt to find data that would aid in assessing and understanding the tank battles that took place in North Africa.
Size: 8 1/2" x 11" • over 100 b/w photos, 30 line drawings • 256 pp.
ISBN: 0-7643-0226-4 • hard cover • $49.95

Galicia Division: The Waffen-SS 14th grenadier Division 1943-1945. Michael O. Logusz. In 1943/1944 a determined group of young men and women in Galicia volunteered to serve in a combat division destined for eastern front combat. Although initially Galicia's Volunteers would serve in a German sponsored military formation, in actuality the volunteers of the Galicia division wanted to engage all hostile ideologies-both from the east and west-in order to secure a free independent Ukraine.
Size: 6" x 9" • over 50 b/w photographs • 592 pp.
ISBN: 0-7643-0081-4 • hard cover • $35.00

Riding East: The SS Cavalry Brigade in Poland and Russia 1939-1942. Mark C. Yerger. Riding East details the history of the SS Cavalry Brigade. Heavily documented, the detailed text is supplemented by 109 photographs, previously unpublished. In addition, six maps, six Order of Battle charts, complete officer rosters 1939-142, feldpost numbers, details of primary commanders within the Brigade, as well information on the units and commanders it was subordinated to, complete a comprehensive history of the first combat mounted unit of the SS.
Size: 8 1/2" x 11" • b/w photographs, documents • 224 pp.
ISBN: 0-7643-0060-1 • hard cover • $59.95

The Last Knight of Flanders: Remy Schrijnen and his SS-Legion "Flandern"/Sturmbrigade "Langemarck" Comrades on the Eastern Front 1941-1945. Allen Brandt. The Last Knight of Flanders is the story of the Flemish volunteers of Germany's famed Waffen-SS, told by those who were there. The book centers around Remy Schrijnen, the only Flemish volunteer to win the heralded bravery award, the Knight's Cross to the Iron Cross. Presents the varying viewpoints of the war as told through the eyes of Schrijnen and those who surrounded him.
Size: 6" x 9" • over 110 b/w photographs, line illustrations, maps • 272 pp.
ISBN: 0-7643-0588-3 • hardcover • $29.95

Comrades to the End: The 4th SS Panzer-Grenadier Regiment "Der Führer" 1938-1945 - The History of a German-Austrian Fighting Unit. Otto Weidinger, former "DF" unit commander. The complete history of SS-Regiment "Der Führer." In addition to Otto Weidinger, portions of the text were contributed by Georg Keppler (Knight's Cross), Sylvester Stadler (Swords to the Knight's Cross) and Otto Kumm (Swords to the Knights Cross). From its creation from primarily Austrian recruits through the Western Campaign, to the bitter battles in Russia where it was almost completely destroyed, the authors explain the engagements of this famous unit in full detail.
Size: 6" x 9" • over 70 b/w photographs, 30 maps • 384 pp.
ISBN: 0-7643-0593-X • hardcover • $39.95

Images of the Waffen-SS: A Photo Chronicle of Germany's Elite Troops. Mark C. Yerger. Mark C. Yerger, in response to requests from readers of his previous books, this new photo album provides material for the model builder, vehicle enthusiast, memorabilia collector and those interested in SS holders of the Knightís Cross.
Size: 9" x 12" • over 320 b/w photographs • 176 pp.
ISBN: 0-7643-0078-4 • hard cover • $49.95